7

JOHN FARRELL

The Day Without Yesterday

LEMAÎTRE, EINSTEIN, AND THE
BIRTH OF MODERN COSMOLOGY

13 Billion Years

1700 | 1800 | 1900 |

THUNDER'S MOUTH PRESS
NEW YORK

THE DAY WITHOUT YESTERDAY
Lemaître, Einstein, and the Birth of Modern Cosmology

Published by
Thunder's Mouth Press
An Imprint of Avalon Publishing Group Inc.
245 West 17th St., 11th Floor
New York, NY 10011

AVALON
publishing group incorporated

First printing October 2005

Library of Congress Cataloging-in-Publication Data is available.

ISBN 13: 978-1-56025-660-1
ISBN: 1-56025-660-5

9 8 7 6 5 4 3 2 1

Book design by Maria Elias
Printed in the United States of America
Distributed by Publishers Group West

For Carmen

ACKNOWLEDGMENTS

ANY TREATMENT OF the life and work of Georges Lemaître must come to grips with the enormous amount of papers and other material he left at the time of his death in 1966. Many people were of invaluable help in the writing of this book and I cannot thank them enough. Among them are Liliane Moens of the Lemaître Archive at the Institut d'Astronomie et Geophysique G. Lemaître at Catholic University of Louvain in Belgium, where Lemaître's papers are collected; Dominique Lambert, professor of astrophysics and philosophy of science on the faculty of Notre Dame de la Paix University, whose extensive biography of Lemaître, *Un Atome d'Univers,* published in 2000, was a critical source of detail about Lemaître's personal life; Dan Lewis, curator of science and technology at the Huntington Library, Art Collections, and Botanical Gardens in San Marino, California; Professor Herman Bondi, the last surviving member of the trio of Bondi, Hoyle, and Gold, which produced the celebrated steady state theory, for sharing his fond personal memories of his meetings with Lemaître in the early 1950s; the excellent staff of the John G. Wolbach Library at Harvard's Smithsonian Center for Astronomy and Astrophysics: Donna Coletti, Melissa Hilbert, William Graves, Amy Cohen, Maria McEachern, and Barbara Palmer, for their help in tracking down now long-forgotten and obscure papers on Lemaître during his visits to the United States, and for their patience over the many months when several large bundles of books went missing from their shelves; Professor Steve Carlip of University of California–Davis for his patient and thoughtful responses to my questions; David Layzer, professor emeritus of astronomy and astrophysics at Harvard University, for his feedback on the manuscript; Barbara Wolff, librarian at the Albert Einstein Archives, Jewish National & University Library; John Grula, librarian at the Carnegie Observatories in Pasadena, California; my parents for their continuous encouragement and support of my writing career; my editor, John G. H. Oakes, of Thunder's Mouth Press for giving me a great opportunity to write about the long forgotten figure of big bang cosmology; Hilary Hinzmann for his editorial help; my agent, Susan Schulman, for her tireless efforts on my behalf; Mary Farrell for proofreading translated material; and, finally, my wife Carmen and my children for their understanding and generous support during the writing of this book.

Any errors or omissions to be found in the following text are entirely the fault of the author.

CONTENTS

If I had to ask a question of the infallible oracle . . . I think I should choose this: "Has the universe ever been at rest, or did the expansion start from the beginning?" But, I think, I would ask the oracle not to give the answer, in order that a subsequent generation would not be deprived of the pleasure of searching for and of finding the solution.

Georges Lemaître, during a discussion of cosmology before the British Association for the Advancement of Science

1.

SOLVAY

We can compare space-time to an open, conic cup. . . .
The bottom of the cup is the origin of atomic disinte-
gration; it is the first instant at the bottom of space-time,
the now which has no yesterday because, yesterday, there
was no space.

Georges Lemaître, The Primeval Atom

NO ONE KNOWS for certain the exact day of that week in
October 1927 when Albert Einstein ran into the round-faced
Catholic priest. If it was Wednesday, then it was a few days
after the new moon. The weather in Brussels may have been
raw, and the forty-eight-year-old Einstein was probably in no

mood to discuss his work with strangers, least of all strangers wearing Roman collars. But that is what happened.

The Fifth International Solvay Conference on Physics ran from Monday, October 24, through Saturday, October 29. Over a week's time Einstein would have ample opportunity to meet the many other physicists who had come to the Belgian capital. On Wednesday, October 26, for example, Einstein could probably be found in one of the meeting rooms of the Institute of Physiology in Brussels, or perhaps not too far from there at the Hotel Metropole, a plush resort built by the wealthy Wielmans-Ceupens family just before the turn of the century. The spacious rooms were an admirable retreat for scientists spending most of their mornings and afternoons in drab lecture halls.

In the event, Einstein was strolling in the alleys of Leopold Park one day during the conference with a former student of his, August Piccard. Piccard, a tall, dark-haired man with a broad forehead, can be seen standing behind the other members in a familiar picture of Solvay taken during the 1927 conference and reproduced in many Einstein biographies. Piccard had written his PhD thesis at the University of Berlin, and Einstein had acted as principal co-examiner of the paper for him.[1]

Piccard was happy to take advantage of the conference to see his old mentor again. The last Solvay conference had been held in Brussels in 1924, just three years earlier. But Einstein had boycotted it in solidarity with the German scientists who had not been allowed to attend (so soon after the end of hostilities, and with feelings on the Belgian and French side still being raw). Now he was here to listen to many of the most illustrious physicists of the day discussing the new quantum theory in papers presented under the conference topic "Electrons and Photons." As common as this topic might sound, electrons and photons were very new objects to scientists in the first quarter of the twentieth century. And the model of the subatomic realm that they were constructing to accommodate these particles was becoming controversial.

Einstein himself was not delivering a paper on that subject or any other, in spite of being invited to do so by the conference chairman (and his close friend), the Dutch physicist Hendrik Antoon Lorentz. Indeed, it seemed to everyone attending the conference that Einstein's role in the proceedings was unusually passive; he was apparently allowing the direction of the development of quantum mechanics to be guided by the Danish physicist Niels Bohr, the young German Werner Heisenberg, and other enthusiastic proponents of this new branch of physics. For many of his friends and contemporaries, such as Max Born and Bohr, this seeming resignation on Einstein's part was a disappointment, but not an entirely unexpected one.

During the past few years, Einstein had grown increasingly uneasy with quantum mechanics, a branch of physics that he had helped to discover in 1905 when he published the four papers that made him famous. In particular, his paper on the "photo-electric effect" described light in terms of little energy packets or particles, called "photons." Up until that time, physicists had accepted the nineteenth-century view that electromagnetic radiation traveled in waves, like ripples across the surface of a pond. They had also assumed that energy radiated in a continuous spectrum, from hot to cold. They were puzzled when in 1900 Max Planck revealed that it did not—energy radiated in very discrete bundles, which, for lack of a better word, the German physicist called "quanta." In 1905 Einstein made use of this new model to describe the behavior of light dislodging electrons from a metal surface the way a cue ball scatters a set of balls on a pool table.

But that was well over twenty years ago now. Einstein had little idea then that his paper on the photoelectric effect would become one building block of a new physics that would describe the entire subatomic world in terms of particles and waves, without ever choosing one definitively over the other as the ultimate model of reality. This new physics would also settle for a description of the subatomic world that depended on the laws of probability and chance rather than strictly determined laws of mechanics, such as those governing the mechanics of Isaac Newton. Einstein felt he could not accept this subjection of atomic physics to statistics and chance, and made one of his most famous quotes in sizing up his feelings when he said, "God does not play dice with the universe."

By 1927 Einstein had come to a watershed. He had been working for some time on a unified field theory, one that he hoped could explain both the laws of gravitation and the laws of electrodynamics in a grand cohesive design. He was willing to grant that the statistical approach to describing quantum interactions was definitely useful, but he did not accept Heisenberg's principle of uncertainty, which the young German physicist had published that year with the approval of Niels Bohr. Nor did he accept Bohr's principle of complementarity and what the Danish physicist declared that quantum physics implied about the world.

Heisenberg insisted that the principle of uncertainty was fundamental in nature. It was not merely a limitation of scientific instruments that prevented the accurate, objective determination of a particular electron's momentum or position, he argued. There simply wasn't any objective momentum or position that existed for that particle in principle—until it was measured. To Einstein, this was rather like saying there was no such thing as a thread in the fabric of your coat until you dug a needle into the texture and tore one out to see what it looked like under a magnifying glass. Suddenly the scientist was inextricably bound up with the very target of his or her experiment, and the results of this experiment depended more on what the experimenter wanted to measure than any objective target in and of itself. This struck Einstein as absurd.

But Bohr carried this implied subjectivism even further with his principle of complementarity. He argued that the subatomic world could be discussed in dual terms—that of waves or of particles—but that an experiment to reveal one such nature of a subatomic system would not also reveal its other nature. Einstein's work on the photoelectric effect indeed described light in terms of discrete particles, photons, but Erwin Schrödinger had developed a model of quantum physics just as reliable as Heisenberg's by describing energy and matter in terms of waves. And it worked. Bohr argued that *both approaches were equally valid.* The quantum world was one of both waves and particles. To Einstein, these statements were more than scientific principles. They amounted to sheer philosophizing as far as he was concerned and simply could not be accepted as the final word on the physics of atomic processes.

Although no one knew it at the time, this 1927 Solvay conference was to be the opening salvo of a number of ongoing and heated debates between Albert Einstein and Niels Bohr on the subject of determinism in quantum physics. The debates, always friendly but intense, would become famous and form the subject matter of many books.[2] In the eyes of most of his peers, Einstein ultimately lost those debates. Einstein maintained that the world was an objective reality independent of the thoughts and attitudes of scientists and everyone else living in it. This reality had to be determinable, and there had to be a scientific model that would completely describe the microcosmic world of the atom as thoroughly and objectively as his own **general theory of relativity** or the mechanics of Isaac Newton described the macrocosmic world of gravitation. But to his discomfort, Einstein soon realized that his general theory of relativity was beginning to evince its own set of troubling characteristics— characteristics he had not anticipated. That was another part of the problem for the physicist at Solvay, part of the reason for his preoccupation: his own grand theory of gravity was becoming unruly. The theory of general relativity was beginning to suggest disturbing ramifications about cosmology, the study of the universe as a whole, and these matters were also begging for attention—an attention Einstein did not feel motivated to give. In spite of his attempts to remain light-hearted at the conference, and several friends reported later that outside of the presentations of papers the great physicist was more his outgoing self,[3] the nagging sense nevertheless lingered for him: that he had made a mistake somewhere in his equations for general relativity.

As a system, general relativity is at its heart a geometric theory, a theory that describes the entire world in terms of geometry. Unlike the classic Euclidean geometry that children learn in grade school, that of parallel lines that never meet and triangles whose sides can never add up to more than 180 degrees, the geometry of Einstein's general relativity is the non-Euclidean geometry derived by the nineteenth-century German mathematician Bernard Riemann. It is the geometry of the sphere, the geometry of curved space. In Einstein's theory, the curvature of space (or, to be more precise, the curvature of **space-time,** the combination of them required by the theory) is determined by the presence of matter and energy. This is radically different from Newton's classical theory of gravity as a universal force that depended on mass alone. In Einstein's theory, objects move along curved paths determined by the presence of matter and energy; as physicist

John Archibald Wheeler once wrote, matter tells space-time how to curve, and space-time tells matter how to move.

Once Einstein had completed the system of field equations for his general theory of relativity, he realized they could apply to the world as a whole. The theory had certain cosmological consequences, once you made a few key assumptions, and it was possible to construct consistent world models using the equations. Einstein himself tried this for the first time in 1917. But other physicists soon followed, and some of the models they suggested using his theory did not only not appeal to Einstein, but they also seriously prompted him to reconsider whether he had a made a fundamental mistake somewhere in his system.

But where? More than likely, he was coming to realize, it was with what he referred to as the **cosmological constant.** Denoted by the Greek letter **lambda,** Λ, Einstein invented the cosmological constant as a sort of buttress, a number he had insisted on inserting into his field equations for gravity in order to construct a model of the world as a closed system, spherical, and in perfect equilibrium—*not changing over time.* This was the way people of the early twentieth century thought of the universe, as a placid, unchanging system. Nevertheless, Einstein noticed early on that his pure equations would not support such a model without some help. The pure equations of general relativity suggested at the outset that a model of the universe would be unstable. The universe of matter and energy would collapse in on itself under the weight of its own gravitational mass, like a balloon quickly running out of air. Obviously this had not happened, the universe was not collapsing according to data from astronomers at the time. So, for the sake of his theory, Einstein felt compelled to offset any possible instability by reinforcing his solutions with a cosmological constant, a factor that acted as a cosmic repellant to the force of gravity, a sort of antigravity that stabilized the world on the large scale of the universe as a whole. It does not appear from available correspondence that it occurred to Einstein that instead of collapsing, the universe his equations described might be expanding, like a balloon filling up with air. In any case, in order to preserve this static equilibrium in his model of the universe (which at the time he thought was the only solution), he had to prop it up with Λ.

That was not the end of his worries, however. Shortly after publishing his cosmological solution in 1917 Dutch astronomer Willem de Sitter examined Einstein's equations and raised another problem, one that would lead to a long controversy between the two colleagues. Among other issues, de Sitter suggested that Einstein's theory could support a spatially flat model of the universe that could be completely empty, without any matter whatsoever. Einstein challenged this view, publishing several papers in response as de Sitter countered with his own. Such a universe could not be possible in his theory, according to Einstein. A universe empty of matter? It would fly in the face of one of Einstein's most cherished principles: that matter itself is what defines the curvature of space-time. How could there be such a thing as space-time that included no matter whatsoever? De Sitter's solutions were mathematically sound, however, and Einstein realized his argument was more philosophical than scientific; he admitted as much to de Sitter by 1920. Neither was he much more confident of this philosophy underlying general relativity by the time the Solvay conference convened in that October of 1927.

It was in Brussels, then, in the muddle of these preoccupations in the alleys of the Leopold Park that Georges Lemaître found him. Einstein had heard of Lemaître, of course. As he regarded the thirty-three-year-old diocesan cleric who caught up with him, he remembered the young man's paper. Théophile de Donder, one of the chairmen of the Solvay conference, had in fact showed it to him earlier in the year, probably at Lemaître's request. It was a clever paper, Einstein could see that, but it reminded him of another he had seen some years before: a bothersome article by some Russian mathematician containing disturbing solutions to his equations that not only argued that the universe of general relativity was dynamic—but also that it could actually expand. Einstein was not impressed.

Accounts of this first meeting between the physicist and the Abbé George Lemaître vary.[4] But common to all of them is the older man's supposed brusque dismissal of Lemaître's idea. "Your calculations are correct," Lemaître years later recalled him saying, "but your physics is abominable." It's unlikely Einstein actually said this to be disagreeable. And it certainly was not meant as a personal insult. More likely Einstein applied "abominable" to the concept of expanding space, because it is essentially the reaction he had to the paper written in 1922 by the Russian mathematician when the whole idea was first brought to Einstein's attention. He simply refused to even consider the idea of an expanding universe at that time.

Instead, he told Lemaître that Alexander Friedmann, a Russian mathematician, had already proposed the same idea in 1922, submitting a paper to *Zeitschrift für Physik* called "On the Curvature of Space," using Einstein's equations to describe the first model of an expanding universe. Friedmann had since died in 1925, without ever having followed up on his initial papers. And being a mathematician rather than a physicist, he had offered his ideas purely on theoretical grounds and had not shown the slightest interest in looking for observational data that might support or contradict his theory. After quibbling with some points about Friedmann's mathematics, Einstein was content to let the matter rest without offering to prove that such an expanding universe was not possible, as he believed. No one else picked up on the idea until Lemaître rediscovered it.

But by 1927 Lemaître had gone considerably further with the theory of expanding space than Friedmann or anyone else ever had. He had already spent a year in Cambridge, England, studying general relativity with Sir Arthur Stanley Eddington. Eddington was the first champion of Einstein's general theory of relativity in England. He was also world-famous for finding what was considered the decisive proof of the theory at the time, with his dramatic expedition to the island of Principe. In 1919 he took photographs there of a solar eclipse that revealed the bending of starlight in the vicinity of the sun, a deflection that matched predictions Einstein had made with his equations. In 1924, when Lemaître came to Cambridge, Eddington was much impressed with the latter's work and encouraged him in his research. Lemaître then spent another year in Cambridge, Massachusetts, studying spiral nebulae with the American astronomer Harlow Shapley, who had used the sixty-inch telescope on Mount Wilson to map the Milky Way. Already indications from these distant sources suggested that the far-flung objects seemed to be mysteriously receding in space. Lemaître believed that such a recession of nebulae was at least preliminary evidence that in fact the universe did correspond to an expanding cosmological model. He had specifically written his 1927 paper with the latest astronomical data in mind. No one would appreciate this for another three years. Far from being put off by Einstein's dismissal, the priest believed Einstein simply did not have any of the astronomical data at hand to change his mind.

And when Lemaître accompanied Piccard and Einstein that day to visit Piccard's lab at the University of Brussels, Lemaître brought up the subject of spiral nebulae and recent measurements of their **redshifts** made by American astronomers to bolster his theory about the **expansion of the universe**. According to his later recollection in a radio address, Lemaître said Einstein did not seem to be aware of any of these facts.[5] Thus the man who would become known as the "Father of the

Big Bang" encountered the resistance of the man whose equations made the **big bang** theory possible.

The modern world's comprehension of the universe is one of the most fascinating subjects in the history of science. But the history of modern cosmology is one of constant doubt, second-guessing, obstinacy, missed opportunities, distraction, and outright denial. Einstein would not accept Lemaître's proposal for an expanding model of the universe in 1927. Not until two years later, when the American astronomer Edwin Hubble himself tentatively published findings that confirmed that most galaxies indeed seemed to be receding from one another, would Einstein relent of this stubbornness, and it would be another two years before he publicly said so.

But the signpost to the expanding universe was already there in Einstein's own equations in 1917. And he knew it. As John Gribbin points out in his book *In Search of the Big Bang* (1986), had Einstein followed where his equations were prompting him to go, he could have specifically predicted the expansion of the universe more than a decade before evidence for it was actually discovered. This would have been the greatest single prediction in the history of science.

2.

OUT OF THE TRENCHES

GEORGES EDOUARD LEMAÎTRE, sometimes referred to as the "Father of the Big Bang," and sometimes less politely as the "Big Bang Man," was born on July 14, 1894, in the Belgian town of Charleroi, about 173 miles northeast of Paris. This was almost exactly twenty years before the onset of what would be known as the Great War. Charleroi was a comparatively recent home for the Lemaître family, which had its roots in Courcelles, to the west of the Belgian mining town, where the Lemaîtres had lived for over two centuries. Young Georges came from a line of men and women who worked with their hands—as weavers and coalminers. Lemaître's great-grandfather, Clement, had served in Napoleon's 112th

Lemaître at 8 months, March 1895. Photo courtesy of the Archives Lemaître,
Institut d'Astronomie et de Geophysique Georges Lemaître,
Catholic University Louvain.

Line and, according to family history, was twice wounded. Demobilized after Waterloo, he retained a devoted loyalty to the exiled leader ever after. But his military career continued, as he became a lieutenant in the Belgian Civic Guard on May 29, 1831, as well as a successful merchant. Clement was married three times, indeed marrying two sisters—one after the other died. His son Edward-Severe became managing director of the Bonne-Espérance Society of Collaries before becoming a wood merchant. He in his turn had six children.

The youngest of these, Joseph, was born in October of 1867 and came from the first generation of Lemaîtres to complete a university education. He studied law, and although he became an accredited lawyer, he did not practice at first. Instead he formed his own business and invented a new procedure to "stretch" glass, one that would prefigure the later methods successfully employed to this day by the European Glaverbel glassware company. When he was twenty-six, Joseph Lemaître married a local woman, Marguerite Lannoy, and they started a family, ultimately raising four sons. In spite of his early success, Joseph Lemaître did not last in the business of invention. A fire devastated his factory, and he was forced to change careers in middle age. Because his company had not been insured, he borrowed money from his cousins to pay his debts and to pay his laid-off employees. His reputation for honesty and sense of social justice as a scrupulous employer spread to Brussels. Impressed by the concern he showed for his business and workers, the Society of Brussels invited Joseph to work for them as a legal consultant to their bank, and in October 1910 he moved his family to Belgium's capital to begin a second career.

By this time, Joseph Lemaître's eldest son, Georges, had already exhibited a facility with mathematics, as well as an interest in theology. Neither displeased his parents, as they were devoted Catholics. But as for making any decision about the boy's future, Joseph Lemaître neither encouraged nor discouraged the nine-year-old Georges when he first told his father he wanted to become a priest. The story is noteworthy, as some later accounts, including one by Lemaître's longtime assistant, Odon Godart, suggested Lemaître entered the seminary out of reaction to his brutal experiences in World War I.[6] But it seems the future cosmologist's religious inclination emerged early on. This does not contradict the likelihood that the horrors Lemaître witnessed in the service as an artillery sergeant may have deepened his sense of vocation.

Like many men of his generation, George Lemaître was born in a year that put him on a direct path toward the trenches. Other key figures in the history of twentieth-century cosmology would also find themselves on the battlefield and, had it not been for the war and its aftermath, might have lived to make even more important contributions to the science. For example, German astronomer Karl Schwarzschild, who provided the first exact solution to Einstein's general relativity equations in 1916, died of a war-related illness shortly after returning from the Eastern front. Alexander Friedmann, being Russian, was on the opposite side of the front. He would be the first to show in mathematical terms that cosmological models in Einstein's theory should be dynamic. Friedmann also suffered ill health because of his hardships during the war. He finally succumbed to complications of an illness contracted after he conducted a meteorological survey in a weather balloon as a professor for the new Soviet Petrograd University (later renamed Leningrad) in 1925. Fortunately, the American astronomer Edwin Hubble, who would discover the evidence of the galaxies receding into space as Einstein equations suggested, was spared. He entered the war when the fighting was all but over, serving briefly in the American army in Europe.[7] It's quite possible that had Schwarzschild and Friedmann lived, they, and not Lemaître, would have been credited more with the discovery of the expansion of the universe and the concept of a beginning point in time and space.

At the time of Lemaître's birth, the man whose work would become the central preoccupation of his professional life, Albert Einstein, was a fifteen-year-old high school student in Munich, watching his own father's business slide into the same sort of oblivion that claimed Joseph Lemaître's. But fire was not the culprit in the case of the Elektrotechnishe Fabrik J. Einstein & Cie. After several years of prosperity from installing streetlights in German suburbs and Italian cities, the company of Hermann Einstein and his brother, Jakob, lost a key contract to light the Munich City Center (probably due to anti-Semitic prejudice) in 1893. Unable to meet their overhead, in July of 1894, the month of Lemaître's birth, the Einstein brothers liquidated their company in Germany and founded a new, smaller company in Pavia.[8] Albert, an unhappy student, remained for a while in Munich. He was supposed to stay for three years to complete the course of his studies. But he missed his family and detested his teachers ("the lieutenants" he called them) so much, he was delighted to leave at the first opportunity. This was provided when his stern professor, Dr. Degenhart, told him he would be better off in another school. Einstein was back with his surprised (and worried) parents by the Christmas vacation of 1894.

As for the wider world at this time, at the turn of the twentieth century, the Catholic Church that Lemaître would serve his entire adult life was going through something of a spiritual renewal. After a century of confrontation with the European nations over its papal states, Pope Leo XIII was steering the Church toward a more accommodative attitude to the modern age than his nineteenth-century predecessors, particularly his immediate predecessor, Pope Pius IX. Leo's most famous encyclical, *Rerum Novarum,* supported the rights of workers and trade unions, and he took a more tolerant attitude to democracy and republicanism. But the Pope also took a keen new interest in the natural sciences—in particular, astronomy and philosophy. We can see some sense of this in the experience of the famous English convert John Henry Newman (whom Leo raised to the cardinalate) on April 27, 1879:

> Newman said that he was living in a curious period. He himself had not the slightest doubt that the Catholic Church and her teaching stemmed directly from God. But he also saw quite clearly that in certain circles a spiritual narrowness predominated that was not of God. To this the new Pope, Leo XIII replied: "Away from narrowness!" He spoke of Galileo as a man "to whom experimental philosophy owes its most powerful impulses." For this Pope, the natural sciences—the Italian Volta, the Swede Linnaeus, the Englishman Faraday—"reached as high a degree of nobility and brilliance as we ever see in man." He was as excited over the railway and other means of communication as if they were miracles. He praised technology or rather man and its creator. "What power he displays when through his discoveries he releases this energy, captures

it again and so directs it along the paths he has prepared for it as to give inanimate material movement
and something akin to intelligence. Finally, he puts it
in the place of man and relieves him of his hardest
labor . . . and the Church, this most loving of mothers,
seeing all this happen, has no intention of hindering it
but rather is glad to see it and rejoices over it."[9]

Pope Leo founded a new academy in Rome to revive interest
in the teachings of the Church's most famous philosopher,
Saint Thomas Aquinas (1225–1274). Both of these subjects,
science and theology, would become central interests of
Georges Lemaître. In less than half a century, he would
become the first head of the Pontifical Academy of Sciences.

This is not to say that such expansiveness on the part of the
Church was consistent during Lemaître's lifetime. Unfortunately, it was all too brief at the start. Leo XIII's papacy was
book-ended by two pontificates now remembered more for
their papal triumphalism and centralization of authority than
for anything else. Leo's predecessor Pius IX, for example,
presided over the loss of the Papal States, the last vestige of
the Church's temporal power, to the Italian government. Perhaps in recognition of this loss of temporal power Pius IX
decided to convene the first Vatican Council. During the
council he promulgated the doctrine of papal infallibility,
though in the final vote the scope of its power was considerably less than what Pius would have liked. The infamous Syl-
labus of Errors, now considered a point of embarrassment

among Catholic scholars, was also issued by Pius out of a sense of urgency and sense that Catholics in Europe and America were under siege from the seductive and dangerous notions of freedom of speech, liberalism, and the merits of public education. Leo's successor Pope Pius X, who was later canonized, was even more utterly hostile to modernization—going so far as to harass and persecute Catholic historians and intellectuals in Europe whose scholarship in any way seemed to accommodate liberal tendencies.

As for the world of physics, it too was undergoing a change at the time of Lemaître's birth. Many popular accounts emphasize a certain complacency in the field at the turn of the twentieth century. Quantum mechanics and the theory of relativity are depicted as earthquakes that took by storm a smug generation of physicists, blindly wedded to the deterministic mechanics of Isaac Newton. But this view is simplistic.[10] In fact, far from being in thrall to pure determinism, scientists toward the end of the nineteenth century were questioning the philosophical underpinnings of mechanics. Drawing on the power of the recently developed disciplines of thermodynamics and electrodynamics, many scientists considered, for example, whether in fact there was any such thing as matter in the concrete, billiard-ball analogy they'd been taught, or whether the building blocks of the world, such as atoms, were merely the appearance, the epiphenomena of a deeper underlying reality (something that string theorists are researching today). This certainly doesn't present as evidence of complacency or lack of imagination on the part of the world physics community as it was. Nevertheless, Alfred North Whitehead's dismissive comment that the last twenty-five years of nineteenth-century physics was "one of the dullest stages of thought since the time of the First Crusade" seems, unfortunately, to have become the common view.[11]

In any case, by 1890 the black body problem was leading Max Planck down the road to the first observations of quantum physics. In astronomy, the ether's questionable existence and the planet Mercury's temperamental perihelion advance were puzzles eluding a solid explanation in terms of classical physics. It was assumed, however, that these riddles would be solved based on the bedrock principles.

But if it's exaggerating to ascribe complacency to the world of physics at the end of the nineteenth century, it's not an exaggeration to ascribe increasing militancy to the relations between the European nations, just a few years before they would plunge into the devastating war that would all but destroy Europe. Years before the assassination of Archduke Ferdinand of Austria on June 28, 1914, tensions continuously sparked between an increasingly imperialistic Germany, its partner the Austro-Hungarian empire, and what they regarded as the meddling triumvirate of Britain, France, and an ominous Russia. It was enough that not only Albert Einstein, by nature a pacifist and mistrusting of the military, could feel a sense of impending hostility between these countries. But even the far-flung young American Edwin Hubble, visiting England as a Rhodes scholar in 1911, felt the bristling air of nationalism when he visited Germany (a country he admired very much) and the sense that world-scale violence was not far over the horizon.

It was in the decade leading up to this flashpoint, that Georges Lemaître got his education. At age ten in 1904, he entered the Jesuit High School of the Sacred Heart in Charleroi. According to his biographer Dominique Lambert, over the six years he spent there Lemaître distinguished himself immediately in mathematics, and, during his last two years, in physics and chemistry.[12] He then went on to a Jesuit prep school in Brussels, the College Saint Michel, in 1910 to study mathematics for the entrance exam to the College of Engineering in Louvain. Of Lemaître's many teachers from this period, one Father P. Ernest Verreux stands out, not only because he represented a role model for young Lemaître as a priest and scientist, but also because of his thoughtfulness regarding the boundaries that separated the two disciplines of theology and science, a subject that would remain central to Lemaître's thinking throughout his life. In later years, Lemaître recalled one occasion from the classroom when Verreux gave him pause. After the young Lemaître expressed himself excitedly on a particular passage from the Book of Genesis that to his imagination seemed to suggest a foreshadowing of developments in science, the older priest suggested Lemaître curb his enthusiasm. He shrugged at his pupil's naive excitement. "If there is a connection," Verreux told his pupil, "it's a coincidence, and of no importance. And if you should prove to me that it exists, I would consider it unfortunate. It will merely encourage more thoughtless people to imagine that the

Bible teaches infallible science, whereas the most we can say is that occasionally one of the prophets made a correct scientific guess." This careful distinction would not be lost on Lemaître, especially decades later when Pope Pius XII ignited a small controversy by reading a little more into the physics of the big bang than Lemaître felt he should have.[13]

The military volunteer: Lemaître in 1914. Photo courtesy of the Archives Lemaître, Institut d'Astronomie et de Geophysique Georges Lemaître, Catholic University Louvain.

In 1911, when it came time for Georges to attend college, he chose engineering as his field. At that time, he felt some financial responsibility to his family; finding work in the coal-mining industry as an engineer seemed the best way to support them.[14] By his graduation, however, it was clear to everyone he knew that his facility with physics and math would be less than challenged in a life of engineering for coal mines. Still, Lemaître plowed on with his course of instruction. But war intervened. World War I would ultimately change the direction of his profession and his life.

In August of 1914, Lemaître and his younger brother, Jacques, were planning a cycling tour of Tyrol with a friend when the Germans attacked France. The Reich armies marched into Belgium that very month. Like so many Belgian men, the two brothers felt a call to national duty. They enlisted on the ninth of August in the Belgian Fifth Corps of Volunteers. After receiving some basic instruction in the use of arms, the Lemaître brothers were shipped to one of the six divisions of the Belgian Third army under General Gerard Leman on October 13. By October 18, Lemaître was in the midst of the Battle of the Yser Canal. This dragged on for two months, until the Belgians, driven back to Nieuwpoort by the Germans, prevented their enemies from breaking through to the sea by opening the drainage channels of the canal to sea-water and flooding the land between the canal and the railway, effectively halting the German advance. The Germans never reached the sea for the remainder of the war, but the price in lives for the Belgian army was high.

Little is known of Lemaître's experience in the trenches. Never one to write letters or keep a journal, we do know from interviews with his family members that Lemaître himself saw a great deal of carnage during his four years in uniform. Lemaître's younger colleague Andre Deprit later wrote that after the flooding of the Yser Canal, Lemaître's corps of volunteers was disbanded:

> Lemaître was detailed to the 9th regiment of line infantry (10 October 1914); the assignment entangled him for several days in bloody fightings house to house in the village of Lombartzyde. After the flood plain had been inundated to stop the German infantry, the 9th regiment was placed on the right flank of the Belgian army. On 22 April 1915, Lemaître watched there the awesome debacle caused by the first attack with chlorine gas; the madness of it would never fade from his memory.[15]

Lemaître was eventually transferred out of the infantry to an artillery division. According to family legend, he incurred the wrath of one instructor when he had the gall to point out an error he discovered in the military ballistics manual. Whether this had any adverse effect on Lemaître's standing is not documented, but by the end of the war, his brother Jacques had attained a commission as an auxiliary lieutenant, while his older brother had to content himself with remaining a sergeant, perhaps, as he half-seriously told a journalist later on, because of "bad character."[16] For all that, Lemaître still received a Croix de Guerre for bravery.[17]

Throughout the fighting, Lemaître's comrades marveled at his ability in the rare moments of calm to continue his studies, reading his textbooks. Lemaître especially enjoyed reading up on electrodynamics and the theories of Henri Poincare, one of Einstein's rivals and one of the most distinguished mathematicians of the era. Indeed, whatever studying he got done in the trenches seems to have been sufficient to convince Lemaître at the close of the war, when he returned to academia, that a career in engineering was definitely not going to challenge him. He dropped his course of study for engineering, and instead pursued his PhD in physical and mathematical sciences.

Lemaître's first step to finish a PhD came in July 1919 at Catholic University of Louvain when he passed the supplementary exams to become a candidate in physical and mathematical sciences. Lemaître's thesis was on the approximation of the functions of real variables under the supervision of one of the country's most respected mathematicians, Charles de la Vallée Poussin. Lemaître obtained his PhD summa cum laude in 1920, along with a baccalaureate in Thomist philosophy. At the same time, he decided on his vocation. He entered the House of St. Rombaut as a seminarian in October 1920. Rombaut was an extension of the main seminary for the Archdiocese of Malines, under the bishopric of Cardinal Mercier, who would become one of the young physicist's mentors and sponsors. Indeed, when Lemaître met the cardinal before entering the seminary, he found his superior not only a profound source of inspiration for his vocation, but also encouraging of his growing interest in relativity, which at the time was becoming one of the most exciting areas in science. Lemaître attended lectures in theology at the main seminary. As Cardinal Mercier wished to accelerate the program of study for seminarians whose careers had been interrupted by the war, Lemaître benefited. For one thing, St. Rombaut's was much less formal than the main seminary, and Lemaître and the other "delayed vocation" students could do what they wished with their free time. For Lemaître, it was time he could spend thinking and writing about physics and theology.

It was during his years at the seminary that several of Lemaître's college professors introduced the future cosmologist to the work of Einstein. Lemaître did not have access to courses on relativity, largely because none had yet been incorporated into college curricula. It had been only a few years since Einstein had completed his general theory of relativity. But Lemaître already had extensive training in geometry and differential equations. He taught himself the tensor calculus to master general relativity, relying largely on Sir Arthur Stanley Eddington's early books on the field. Lemaître wasted no time acting on his new favorite subject. In 1922 he wrote his own short thesis, "The Physics of Einstein," which he entered in a competition for a scholarship from the Belgian government to study abroad. Lemaître won the scholarship, which would allow him to travel across the channel to the United Kingdom for a year of study—and to meet the man whose books had opened up a new world to him. As for the rest of his training, Lemaître spent three years in the seminary before being ordained. His biographer Dominique Lambert has written that if Lemaître did not particularly distinguish himself in the study of theology—especially neo-scholastic theology as it was then taught—it in no way dampened or discouraged his sense of vocation. (Indeed to some of his later critics, Lemaître's less than stellar facility with theology, a decidedly rarefied subject especially in Catholicism, would probably go as a mark in his favor as a physicist.) But for Lemaître, the spiritual side of theology appealed to him more than the intellectual side.

It's not clear at this early stage of his interest in relativity whether Lemaître already suspected Einstein's equations suggested an evolving universe. Still, his immediate plan of action was to study not only astronomy but the new field of astrophysics as well. It is clear he was interested from the start in observational data as it applied to general relativity, and by the time his two years abroad in the United Kingdom and the United States were over, Lemaître would have all the inspiration he needed for his life's most important work.

But in 1923 when Lemaître became a priest and planned to go to Britain, the evolving universe was still in the future. The idea of a dynamic cosmos was not one that occurred to anyone of Einstein's generation prior to the emergence of his general theory. Since 1917 Einstein had been aware of the possibility of a dynamic cosmos, based on his own equations. But it was a possibility he did not want to confront, and this is odd given the revolutionary nature of his theory of relativity and its fundamental implications for physics. For all the elasticity that relativity demanded of time and space now, Einstein balked at extending that elasticity to the very cosmos itself. He wanted to keep the universe he grew up with. He wanted to keep the static, stable majestic cosmos of Newton.

But Einstein was not to be indulged in this matter. As we saw in the previous chapter, within months of publishing his paper on the cosmological implications of general relativity, Einstein's friend and colleague Willem de Sitter published papers not only discussing the "instability" of the Einstein universe, but also positing a theoretical universe containing nothing but empty space. Einstein vigorously contested this idea; for him the underlying assumption of his theory was that the shape of space itself is influenced by the presence of matter, therefore there could not be any space without the presence of matter. This "argument" between the two friends went on for some months, with each of them trading volleys in the pages of the physics journals. Indeed, to interested readers the Einstein–de Sitter "controversy" is perhaps more fascinating than the vaunted showdowns between Einstein and Niels Bohr on quantum physics that took place years later.

Einstein found himself even more caught up in controversy when a relatively unknown mathematician named Alexander Friedmann wrote to show him that the cosmological constant could be set to zero (in effect, it could be dispensed with)—if he realized that his equations suggested a universe with a radius that could expand or shrink *over time*.

This was not at all to Einstein's liking. When he saw Friedmann's paper "On the Curvature of Space," published in *Zeitschrift für Phyzik* in 1922, he dashed off a hasty response claiming that Friedmann had made an error in his equations. In fact, Einstein himself was the one who made an error in his haste to correct the Russian. He would later retract his "correction" and concede for the first time that mathematically Friedmann's equations could indeed suggest an expanding universe. But he continued to resist the notion that such a solution actually had any basis in reality. No one else seems to have taken note of Friedmann's notions, and by the time Lemaître independently worked out his own solutions, the Russian had died.

Did Einstein suspect the universe was dynamic back in 1917? Why did he insert lambda into his equations? In his paper "Cosmological Considerations on the General Theory of Relativity," he simply states, "That term is necessary only for the purpose of making possible a quasi-static distribution of matter, as *required by the fact of the small velocities of the stars*" (emphasis added).[19] Or did he realize the possibility of expanding space—and did he lose his nerve?

Einstein's hasty objection and retraction concerning Friedmann's work seem to suggest a stubbornness based on more than simply what he knew of the stellar velocities of the time. They suggest a prejudice, an ingrown view of the cosmos inherited but never really questioned. There never had been reason to question it until the advent of Einstein's theory. His attitude gets right to the heart of the question—just how wedded was Einstein's generation to the notion of a serene static universe?

But before returning to Lemaître and his role in shaking up that static view of the cosmos, it would be helpful first to take a brief detour into history, a survey of the "static" cosmos that so preoccupied Einstein's generation—and indeed the many, many generations that preceded his.

3.

A UNIVERSE THAT EVOLVES:
THE HISTORY OF AN IDEA

Physical scientists have a healthy attitude toward the history
of their subject; by and large we ignore it.

P. J. E. Peebles, "Impact of Lemaître's Ideas on
Modern Cosmology"

IN A SENSE the title for this chapter is a misnomer. Prior to
Einstein's theory and the period when its cosmological consequences were worked out in the 1920s, there really was no
history for the idea of a dynamic cosmos. On the other
hand, Aristotle's ancient notion of a pure, far, and changeless firmament of the heavens, which came to dominate
Western thought right up to the time of the telescope and

Galileo's first fine sketches of craters on the moon, was as far removed from the cosmogonies of other ancient peoples as the dynamic models of Lemaître were from Isaac Newton's physics.

The average reader today thinks nothing of the notion of an expanding universe, one populated with exotic objects like black holes, quasars, and "dark matter," stars that explode in space with an energy many times greater than the output of their own galaxy, and orbs that carve out bottomless pits of nothingness in the very heart of each galaxy. So it may seem a little odd to understand about the period of Einstein and Lemaître that for centuries Europeans had grown up with the idea that the universe was an unchanging firmament. And Europeans had thought this since before the days when they translated Aristotle from the Arabs, who in turn had translated him from the Greek that the Europeans had long since forgotten.

About a thousand years before Lemaître headed off to England for his post-doctoral work, the elite priests of Mayan culture were applying their mathematical skills to the stars that rose over their Central American empire. The stars and the planets were important to the Mayans. Unlike the ancient Egyptians, for example, and Babylonians (who boasted sophisticated knowledge of astronomy), the Mayans were less interested in the use of accurate observations for agricultural purposes as they were for the keeping of their calendars, or *tzolkin.*[20] The neighboring Inca and Aztec empires were no less devoted to this preoccupation with time. Never-ending cycles of birth, death, and rebirth were the order of the Mayan cosmos. The idea of the world having a history—a beginning, development, and eventual end—was virtually unknown among ancient peoples (the exception being the Hebrews). Time was a system of endless cycles, and the Mayans, like most other ancient folk, wished to inject their own identity as a nation and as individuals into these recurring cycles. This was a way to escape the apparent finality of human death and decay. To help with the observance of rituals designed to give them a sense of belonging to the world of the apparently changeless, they studied as tools the wheeling of the stars during the course of the night, the phases of the moon during the course of the month, the return of the seasons and the revolution of the planets during the course of the year to keep the calendar's relentless measure of time in order.

All of these repeating courses were nestled inside a much larger cycle that was running its course, a cycle that ancient peoples very much wanted to be able to describe—if not contain—for their sense of immortality. For the Aztecs, for example, the world created and destroyed itself four times in the span of 2,500 years. For the Egyptians, it's much less clear, as ancient sources are sparse and considered unreliable, but it was likewise probably on the order of a few thousand years per cycle. For the Chinese, the cosmic period of life and death was called a *yuan,* consisting of twelve periods, or *hui,* of 10,800 years each. In Chinese cosmogony, after 129,600 years the world ended and was reborn again. For the Babylonians, the life cycle of the cosmos was broken down to six hundred periods known as *saros,* each of them being 3,600 years long, making the entire span of the cosmic cycle an impressive 2,160,000 years.

For the ancient Hindus of India, however, the cycle was magnitudes larger. Indeed, Ancient India distinguishes itself as the only culture before the modern period that grappled with the idea of eons comparable to the huge numbers astronomers routinely deal with today. In Hindu cosmogony the world was believed to transform itself every 4.3 million years. But this cycle was itself one embedded in an even larger age of ages—4.3 *billion* years long before its death and regeneration. The Hindus believed in these massive cycles of millennia passing from one great age of the world to the next, coming full circle only after eons of change. The classic Puranas, a collection of sacred writings from the middle of the first millennium, put the Hindu desire for immortality in perspective:

> I have known the dreadful dissolution of the universe. I have seen all perish, again and again, at the end of every cycle. At that terrible time, every single atom dissolves into the primal, pure water of eternity, whence originally all arose. Everything then goes back into the fathomless, wild infinity of the ocean, which is covered with utter darkness and is empty of every sign of animate being. Ah, who will count the universes that have passed away, or the creations that have risen afresh, again and again, from the formless abyss of the vast

waters? Who will number the passing ages of the world, as they follow each other endlessly?[21]

Beside all these ancient cosmogonies, based on myths of the world culled from chaos by the acts of Gods, the ancient Greek view of the world, as one subject to laws of nature that could be investigated rationally, seems eccentric. No less eccentric was the view, adopted by Aristotle, that the heavens were a crystalline structure of shells within shells in which the planets and stars wheeled around the earth for eternity—unchanging, with no beginning, no ending. The heavens were pure, according to Plato and Aristotle, pure as a world set apart from the crude matter of the earth. The heavens had always existed, according to Aristotle, and always would. Above all, in contrast to the universes of the other ancient peoples, the universe the Greeks believed in was composed completely of inanimate matter. It was not considered the phantasmagoric excretion of warring gods. This is not to say that the Greeks didn't have their gods; they of course make up a familiar pantheon. But the Greeks were unique in distinguishing between their deities and the physical world in which they took part. Yet, while the Greeks did not believe the planets were themselves animated beings, as did the Babylonians and Egyptians, neither did they believe their composition resembled anything similar to what existed on earth. The matter of the heavens was considered separate from the matter of the earth. It was Galileo's telescope that first revealed that sun, moon, and planets had the pockmarks of common earthlike matter.

This static, eternally unchanging cosmos of the Greeks is the one that passed down through the centuries to the West, for all the modifications of Copernicus and Galileo and Newton, right to the nineteenth and early twentieth centuries, when Einstein began to assemble his grand theory of relativity. Gone were Aristotle's crystalline spheres, to be sure, and gone was the notion that the planets and stars were made of some ethereal unearthly matter. But otherwise the vast heavens in which asteroids, meteors, planets, stars, and nebulae drifted were still considered by scientists of the time as changeless.

While it would be inaccurate to describe the other ancient cosmogonies as "static" in the sense that the crystalline spheres of Aristotle can be regarded, one cannot really attribute the dynamics of a truly evolutionary view to the traditions of the Hindus, the Chinese, or the Mayans either, given the inherently cyclical nature of time as they perceived it. Further, while in these views the substance of the cosmos was not literally motionless, but going through repeating cycles of birth, death, and rebirth, what the other ancient cosmogonies lacked was any quantifiable sense of how to measure the change over time. Only in the first third of the twentieth century would relativistic theory find a home for cyclical models of time and space in cosmology. Nevertheless, one can glimpse in the eternal cycles of the Hindu cosmos, and that of the Mayans and Babylonians, a cyclical world similar to offshoots of the later oscillating relativistic models of the universe that forever expanded from a singularity, contracted into collapse, and expanded again into a new universe. And in terms of the billions of years between each cycle, as noted above, the ancient Hindus were not far off.

Tempting as it is to view science through the ages as a steady march of progress from the Greeks up to the present day, the stultifying effect of certain stubborn principles hindered progress in astronomy and cosmology again and again over the centuries, even as the mathematical tools available to astronomers improved their calculations and predictions. For example, the concept of perfectly circular motion, an idea the Greeks espoused at least since Pythagoras in the sixth century BC, was regarded as an eternal truth of mathematics and, by extension, of the motion of the planets as well. Useful as this idea may have been in the development of Greek geometry, it played havoc with astronomers from Ptolemy right up until the time of Johannes Kepler in the seventeenth century. Even Copernicus, with his revolutionary idea of a sun-centered solar system, could not escape the influence of purely circular orbits.

It was clear from observing the planets, Mars in particular, that they did not follow perfectly circular orbits. And yet the dogmatic attachment to a purely aesthetic principle of circular motion muddied Copernicus's theory, forcing his continued reliance on cumbersome epicycles, which he added to each of the orbits of the planets in order to bring his system into agreement with the actual observations of the planets' positions. And even Kepler hesitated before adopting the correct but "ugly" idea that the planets actually orbited the sun in ellipses, with the sun as one foci.

In a similar fashion, blind acceptance of Aristotle's theory of motion, a theory of impetus, retarded the development of a fruitful system of mechanics for hundreds of years until the thirteenth- and fourteenth-century monks of the University of Paris tossed it out (for largely religious reasons) in favor of a new principle. This principle would eventually become the law of inertia, Newton's First Law and the foundation of modern mechanics: that an object will continue in its path unless acted on by other forces.

This is a fascinating and often overlooked crossroads in the history of science. In essence, Jean Buridan and Nicholas Oresme, both professors at the University of Paris, challenged Aristotle's essentially useless theory of impetus when applied to heavenly bodies. Buridan (1300–1358), a disciple of William of Occam, taught scholastic philosophy. His pupil Nicholas Oresme (1323–1382) taught theology—but both thought deeply about Aristotle's physics (Oresme developed a system of coordinate geometry in advance of Descartes).

Aristotle argued that an object needed constant attention from a source of motion in order to keep moving. Thus an arrow loosed by an archer could travel for a great distance due to the added impetus given to it by the air supposedly falling in behind it as it moved, perpetuating its momentum. In the same fashion, Aristotle suggested, the planets revolved around the earth due to the constant force exerted on them by a "prime mover." To the medieval monks of Paris, this sounded like nonsense. Specifically, they questioned why God should have to continuously intervene to impart motion to one of the planets when a simple impetus of motion from the start of the cosmos should have been sufficient to last for the duration of the universe. What may seem today to sound like a theological quibble actually cleared the cobwebs of Greek preconceptions out of the European mind, allowing a more abstract approach to the theory of motion, so that by the time of Leonardo da Vinci and Copernicus, the idea of inertial motion, imparted by and depending only on an initial force, was virtually taken for granted by astronomers.

In the seventeenth century, Isaac Newton would establish inertial motion as the cornerstone of his three laws of mechanics: *Every body continues in its state of rest, or of uniform motion in a right [straight] line, unless it is compelled to change that state by forces impressed upon it.* By the time of Newton's death, his laws of mechanics, his law of universal gravitation, in combination with Kepler's laws of planetary motion and Galileo's introduction of the telescope, armed European scientists with the tools they needed to accelerate the exploration of the cosmos and the development of modern cosmology.

It is tempting to speculate that as early as the eighteenth century perhaps Newton himself or some contemporary scientist building on his theory of universal gravitation might have apprehended that the universe in some fashion needed to be evolving in order for Newtonian theory to be consistent. After all, Newton himself was confronted with the consequence of universal gravitation right from the start—if gravitation was truly universal, what was to prevent the cosmos from collapsing under its own weight? This was the question Einstein had to face and that inspired his use of the cosmological constant. Newton neatly dispensed with the inconvenience of this question by supposing that the universe must be infinite in size and number of stars, and that this would prevent any possibility of collapse. Later critics would discover that this elaboration would not in fact work—that in mathematical terms, Newton's theory of universal gravitation could not be spared the cosmological consequence of cosmic collapse by assuming infinite space and infinite stars.

Astronomers at the time, of course, had no means of recording their observations on photographic plates or detecting the actual composition or velocity of stars and nebulae with spectrographs. Even if such methods had been available, Newton's theory was not at its core a geometric theory; it did not lend itself to cosmological considerations the way that Einstein's theory would. Immanuel Kant would go so far as to question whether the universe as a totality of existing things even had any physical meaning outside of the human mind.

For purposes of the present discussion, the history of science's outgrowing the view of a static, nondynamic cosmos, and the foreshadowing of twentieth-century cosmology, begins with two philosophers—Thomas Wright and Immanuel Kant—in the eighteenth century. Wright, born in England in 1711, was largely self-taught and made his living as a tutor to aristocrats. In 1750 he published *An Original Theory or New Hypothesis of the Universe,* which was largely concerned with the theological issue of where the center of creation might exist. Toward the end of the book, Wright suggested that the universe was a disk-shaped conglomeration of stars that we could discern only from the side, since the Earth and the solar system were embedded in an outer arm of the disk. He did not go into this theory in any detail, and it was left to the young Kant, long before the writing of his *Critique of Pure Reason* and his doubting of the universe, to publicize this view in more depth. In *General History of Nature and Theory of the Heavens* in 1755, Kant speculated that not only was the Milky Way a disk of stars all moving together in the same plane, but that the fuzzy nebulae that astronomers were beginning to catalog as distinct from the stars were in fact conglomerations of stars that were too far distant to discern individually. In essence, he argued, these nebulae represented separate "island universes" in their own right.

A countryman of Kant's, William Herschel, would expand on this idea with his own discoveries. Born in Hanover in 1738, Herschel began his career following in his father's footsteps as a musician. When the French occupied his hometown, he moved to England, where he eventually settled for a time as the organist at the Octagon Chapel in the seaside town of Bath. During this time his interest in astronomy, at first merely a hobby, became an obsession to the point that Herschel learned to build his own telescopes. With the help of his sister, Caroline, he began detailed observations of the stars and planets, hoping to discover a new comet. In those days, discovering comets was considered a sure path to fame and fortune. Instead of discovering a comet, however, Herschel discovered a new planet, Uranus, in 1781.

Herschel was appointed court astronomer by a pleased King George III when he offered to name the planet Georgium Sidus (Star of George) after the monarch. (The planet was not known as Uranus until the nineteenth century.) With support and patronage from the king, Herschel could afford to build bigger and better telescopes, and he determined to improve upon the catalogs of nebulae that French astronomer Charles Messier had begun in 1760. Taking his cue from Kant, Herschel also argued that nebulae, many of which he was able to discern as conglomerations of distinct stars, were island universes of their own. But he went further and argued that such universes, like the Milky Way itself, coalesced under the force of gravity into their current structures. In essence, Herschel became one of the first to describe the universe in evolutionary terms. By 1820 Herschel had catalogued over two thousand nebulae, but his contention that nebulae represented separate island universes became complicated by the fact that he could not discern individual stars in very many of them. Too many seemed to be merely balls of bright gas. For this reason, the idea of separate galaxies remained just an idea for another century until Edwin Hubble was able to determine a relatively accurate distance to the Andromeda Nebula in 1925.

Whether Herschel's work had any influence on the young American writer Edgar Allan Poe is not known. But not long after Herschel's death in 1822, Poe, departing from his gothic tales of horror, wrote a fascinating discourse on the universe, titled *Eureka*. In no sense to be taken as a scientific treatise, according to Poe, the work was written simply to exercise his imagination. But Poe's imagination served him well and has often given him a footnote in histories of cosmology. He suggested the entire universe had developed originally from an explosive ball of fire, expanding outward until the stars coalesced out of its dispersion and slowly evolved into the Milky Way.

As inspiring as Poe's essay was, science did not have to wait long for a new discovery to change the way astronomers viewed the heavens and eventually give them the means of testing Poe's flight of fancy. This was the development of **spectroscopy,** an offshoot of Isaac Newton's experiments with optics, when he beamed white light through a prism to separate the individual colors, or wavelengths, into the rainbow that makes up the visible **spectrum**. An exhaustive study of spectroscopy as it developed in the nineteenth century could easily fill several books. But three key developments that directly affected the development of modern cosmology are worth mention. The first was the invention of the spectrometer by Joseph Fraunhofer in Germany in 1814. With this new device, which could more accurately display the spectrum of visible light, Fraunhofer compared the spectra of flames in the laboratory with the spectra of the sun and even the stars. He noticed that the spectra for the sun and the stars evinced certain dark lines, very similar to those manifested by the spectra of flames he had produced in the lab from ordinary elements. And soon other scientists, notably

Herschel, suggested that spectroscopy could be used for chemical analysis.

The significance of the spectra was established by Gustav Kirchoff and Robert Bunsen (inventor of the Bunsen burner) in 1860. Bunsen examined the spectra of flames lit from different elements. They gave off their own rainbows with characteristic absorption lines, dark bands in the spectrum that served as physical fingerprints. Different chemicals, such as carbon, showed different absorption lines. The absorption lines represented the discrete wavelengths of visible light that were absorbed by the chemical and not passed on to the spectrum. By experimenting with several metals, Kirchoff and Bunsen showed that each substance had its own line spectrum. Bunsen went further, analyzing the spectra of alkaline compounds, and soon discovered the existence of two new elements—cesium and rubidium. By analyzing the spectra of sunlight and comparing it to the spectra of basic elements such as sodium, Kirchoff realized that a substance that could emit a spectral line, or emit light of a particular wavelength, could correspondingly absorb it as well. He realized the spectra of the sun were telling him that the sun itself contained elements of sodium, carbon, and many other elements. Thus in 1860, for the first time, scientists realized they could study the chemical composition of stars. The field of astrophysics was born.

Before the end of the decade, astronomer William Huggins in England had examined the spectra of several stars. He noticed that one of the brightest, Sirius, had a spectrum with

Lemaître, circa 1920, just prior to entering the seminary. Photo courtesy of the Archives Lemaître, Institut d'Astronomie et de Geophysique Georges Lemaître, Catholic University Louvain.

absorption lines that, in comparison to spectra from light taken in the lab, seemed displaced toward the red end of the rainbow spectrum. It was Huggins's insight to realize that this displacement was due to a phenomenon first suggested by Christian Doppler in 1841 and ever after called the **Doppler effect.** Doppler suggested that movement in a source relative to an observer would manifest itself in a change in wavelength from that source. If the source was a source of sound, the change in wavelength amounted to a change in pitch (as when an ambulance siren goes blaring by); if the source was a source of light, the change in wavelength amounted to a change in color. Based on his analysis, Huggins determined that Sirius, the "dog star," was moving away from the sun at a velocity of 29.4 miles per second. This discovery would have momentous consequences for Lemaître's work.

As spectroscopy was developed, another branch of physics was also evolving, the field of thermodynamics, the dynamics of heat energy. This field would also inspire scientists to break free of the age-old acceptance of the changeless universe, albeit in a somewhat negative fashion. The second law of thermodynamics, in particular, the law of entropy, disturbed not only scientists but philosophers as well. If the total energy of any system was doomed to run down, dissipate to its lowest, most useless form, then clearly the universe itself was ultimately doomed. Scientists began talking of the "heat death" of the universe. In the face of this grim eventuality, English physicist William Thomson, Lord Kelvin, was the first to apply the law of entropy to the universe as a whole. Realizing that entropy must be increasing, he suggested it should be possible in theory to "work backward" to an earlier epoch of the cosmos by using the laws of thermodynamics. In a sense, Kelvin was applying the reverse logic of thermodynamics in the same way that Lemaître would later apply reverse logic to the field equations of general relativity. Kelvin's suggestion was entirely qualitative; he did not offer any systematic application of thermodynamics to the universe as a whole; it was just a speculation and nothing serious was made of it (meaning, no one thought of a way to test his idea). But his musings were another foreshadowing of the change scientists were undergoing in their attitude to the universe and its evolution.

By the very early twentieth century, astronomical observations that would accelerate that change were already being made. In 1912, when Lemaître was a student of eighteen, American astronomer Vesto Slipher used the Lowell observatory in Flagstaff, Arizona, to begin taking the spectra of nebulae. This was a painstaking process, and it took him a few years before he had completed taking the spectra of fourteen nebulae. He noticed that all but two of them were shifted to the red end of the spectrum. By now, it was generally accepted that this shift was due to the Doppler effect and suggested a velocity of recession. But with only fourteen nebulae measured in this fashion, nothing definitive could be said.

Once Edwin Hubble measured the distance to the Andromeda Nebula in 1925, astronomers realized the Milky Way was indeed as Kant and Herschel thought, just one of many galaxies in a larger cosmos. But the universe was still considered static—or rather it had not occurred to anyone but Friedmann, himself only a mathematician, and Lemaître, still working toward his PhD, that the universe might change in size with time. Friedmann died before the year was out. Lemaître was still groping his way toward a truly dynamic view of the cosmos.

This admittedly brief survey over the vast expanse of history has been adopted to show the relatively unchanging view of the cosmos that prevailed in the West from the Greeks through the time of Einstein and how it finally began to change. One could argue that until Einstein there was no consensus among thinkers that the universe per se was even a coherent graspable object of study.[22] It seemed an endless void of stars; indeed, according to Newton it had to be. How else explain why they universe had not collapsed under the weight of universal gravitation? And as radical as Einstein's theories were about the laws of electrodynamics and energy quanta, he too accepted the general view of an unchanging and essentially eternal universe. It was a universe as indeed static as a painting of the crystalline spheres posited by Aristotle and the ancient Greeks.

And it was by no means a symmetrically shaped universe. In 1917, when Einstein first applied his theory to the question of the universe, the Milky Way was considered to be all there was. The Andromeda Nebula, now recognized as our closest galactic neighbor, was still considered a small cloud of stars and dust that floated above the general plain of the Milky Way. There were no other "island universes" outside our galaxy. For all practical intents and purposes, the universe was the Milky Way, and the Milky Way was the universe. The methods that were devised to get a better sense of interstellar distances were only slowly being applied, so it would be a few more years before Edwin Hubble determined the true distance to Andromeda and established that it was a separate universe unto itself.

It's somewhat odd to see in retrospect how conservatism and hesitation manifest themselves in the history of science— sometimes, as we saw in some elements of Greek thought, for centuries. Perhaps because of the general nature of this conservatism, it's unfair to ask why Einstein was content to plug the cosmological constant into his equations—in reaction to their clear suggestion that the universe as a whole could not be static. Einstein was preceded by generation after generation of thinkers who believed the same thing. Einstein went by the data he knew at the time, although its clear from Lemaître's recollection of their meeting in Brussels in 1927 that he wasn't that alert to the most recent studies and measurements. It would be up to Lemaître now to lead Einstein and the rest of his generation into a new, truly dynamic model of the universe.

4.

FROM CAMBRIDGE
TO CAMBRIDGE

IN OCTOBER 1923 Lemaître crossed the English Channel and arrived at Saint Edmund's House, Cambridge University's residence for Catholic clergymen. Sir Arthur Stanley Eddington had accepted him for a year's study at the observatory as a research student in astronomy, and the young priest enthusiastically jumped into his studies. In addition to regularly attending Eddington's lectures, Lemaître also attended those by Ernest Rutherford, one of the fathers of atomic physics, who first posited the existence of the nucleus and explained the underlying dynamics of radioactivity. Apparently, Rutherford was the better lecturer.[23] Nevertheless, it would be difficult to overstate the importance of

Eddington's influence and inspiration for Lemaître. Short of being able to study directly with Einstein himself, Lemaître landed the next best authority on the subject at the time, the man whose explications of Einstein's theory were Lemaître's first introduction to the field. Eddington was the first champion of Einstein's theory for the English-speaking world. In 1916, at the height of World War I, the Dutch astronomer Willem de Sitter sent Eddington Einstein's principle papers on relativity (Holland being a neutral country and open to communications). As noted in chapter 1, de Sitter was one of Einstein's friends and among the very first to begin applying the equations of general relativity to cosmology. Eddington immersed himself in the new theory and immediately understood its importance for astronomy and astrophysics.

By 1919, with the war still dragging on, Eddington had written a concise summary of the theories for the English-language physicist, later publishing a longer overview, *The Mathematical Theory of Relativity*, in 1923 (which Lemaître had read). But most dramatically, in 1919 Eddington personally led the famous eclipse expedition to the island of Principe off the coast of the West African nation of Guinea. Eddington photographed the sun during its occultation by the moon and confirmed one of the three key predictions made by Einstein's theory: the bending of light by a gravitational field.[24] Eddington's *Internal Constitution of the Stars* remains a classic in the field of astrophysics. While quantum physics was still in its developmental stage, Eddington turned his attention to stellar evolution, trying to ascertain how quantum processes might explain how stars generate energy. Eddington established a link between a star's mass and its luminosity and generated some controversy, granting ages to stars many times longer (in the billions of years) than scientific opinion at the time had allowed for. His interest in relativity, particularly the mass/energy relation of $E = mc^2$, led to his development of some explanations of how stars might transform matter into energy. Not all of his theory turned out to be accurate, but some of his work would be incorporated into the Nobel Prize–winning theory of Hans Bethe, who in 1938 developed the first detailed explanation of how nuclear processes inside the stars fuse hydrogen into helium. (Bethe would later show up as a fictitious coauthor of one of the most famous papers on big bang theory in 1948.)

Eddington had been impressed by Lemaître's scholarship-winning paper "The Physics of Einstein" and appreciated the young Belgian's aptitude for the field equations of general relativity. There were no classes in those days in tensor calculus or differential geometry, as noted in chapter 2, so Lemaître essentially taught himself the mathematics, using Eddington's text to master the theory on his own. Eddington was also, apparently, impressed by Lemaître's collar, donned so soon after his distinguished military service in the war. Eddington was a Quaker and a pacifist. He'd avoided military service as a conscientious objector and barely avoided internment in a camp for the duration of the war by volunteering for the very expedition tour that would make him and Einstein famous.[25] Perhaps Eddington saw in Lemaître, the decorated sergeant and priest, a man very similar to himself: a religious loner, unmarried, outstanding in mathematical ability and with a fresh mind open to possibility. Not only would Eddington be the first to appreciate Lemaître's contributions to the growing corpus of work in general relativity, but he would also provide Lemaître with some of the "problems" that would inspire the Belgian to address old questions and forge new paths in cosmology.

There's no doubt that Lemaître was equally impressed and inspired by his English mentor. In particular, Eddington's balance of physical theory with his interest in astronomy and the comparatively new discipline of astrophysics encouraged Lemaître to build his own theories from the outset with the goal of grounding them in actual observations and data. As we shall see, this is what distinguished Lemaître from virtually all the other specialists interested in general relativity and its cosmological implications at the time.[26]

Eddington was superb at fielding problems for his protégés to crack. And Lemaître would prove to be a master not only at solving the problems Eddington gave him but also at improving upon his mentor's methods. Under Eddington's supervision, Lemaître planned in the 1923–1924 academic year to write a paper on the concept of simultaneity in general relativity: how it must be modified when shifting from objects accelerating in a straight path to objects accelerating in curved space. Specifically, Lemaître argued that events that appear to be simultaneous on a uniformly accelerated rigid body will not necessarily be simultaneous on a uniformly rotating body. This point would be of some import much later when Lemaître addressed the apparent limits of the **Schwarzschild solution,** a solution to the equations of general relativity that would have ramifications for the study of black holes. At once pleased with Lemaître's thesis, Eddington wrote a foreword to Lemaître's paper on simultaneity before sending it off for publication in *Philosophical Magazine.* Shortly after, he described his impressions of his new student to friend and colleague Théophile de Donder, another Belgian physicist:

> I found M. Lemaître a very brilliant student, wonderfully quick and clear-sighted, and of great mathematical ability. He did some excellent work whilst here, which I hope he will publish soon. I hope he will do well with Shapley at Harvard. In case his name is considered for any post in Belgium I would be able to give him my strongest recommendations.[27]

If his travel itinerary can be viewed as any indication of Lemaître's success over the next decade, the priest was at the heart of modern cosmology's development from 1925 right up until the outbreak of World War II in 1939. As previously noted, Lemaître realized the importance of trying to undergird the complicated geometry of general relativity with some evidence of its effects on the grand scale of the cosmos—if he could find them. His interest therefore took him from Cambridge, England, to the Cambridge on the other side of the Atlantic— the Cambridge of Harvard College and Massachusetts Institute of Technology (MIT). With a new scholarship from the CRB Educational Foundation, funded by the Committee in Relief of Belgium, Lemaître now planned to meet and study with some of Eddington's distinguished colleagues—chiefly those who specialized in the direct observations of stars and nebulae with some of the world's best equipment. In those days the world's best telescopes could be found in only one place, the United States. Harlow Shapley, late of the Mount Wilson Observatory and director of Harvard College Observatory from 1920 until 1952, welcomed Lemaître for the academic year of 1924–1925.

Shapley could be an irascible character, stubborn about what he believed—even when he turned out to be wrong. His imposing portrait can be found today inside the lobby of Harvard University's Wolbach Library on the grounds of the observatory. In spite of his modest physical stature, there is something determined about Shapley's wide face and close-cropped hair. In his staid suit and tie he resembles nothing so much as the all-knowing Dr. Van Helsing–style characters of many melodramas from Hollywood's Golden Age. Shapley had eccentricities to go along with this persona. For example, when he was bored examining photographic plates or couldn't sleep after spending a night freezing in the dome of Mount Wilson, he could go out during the day and study local ant populations on the mountainside, going so far as to collect data and submit a paper full of observations on the "Thermokinetics of Dolichoderine Ants" to the *Bulletin of the Ecological Society of America.*

As a fellow native Missourian, the young astronomer Edwin

Hubble regarded Shapley, his first boss, as a rival as soon as he came to Mount Wilson to commence what would become the most dramatic observations of the era. But Hubble was a close, competitive, and guarded man who did not get along well with Shapley, and he was happy when the latter was soon promoted to director of Harvard College Observatory in 1920.

It was thanks to Shapley's work, poring over hundreds of existing plates and taking many new ones of globular clusters, large conglomerations of stars, that astronomers came to realize the actual length of the Milky Way, the location of the galaxy's center, and the solar system's relative place within it. Certain stars, known as **Cepheid variables,** were plentiful in globular clusters, which were long known to clump largely in one hemisphere of the sky. Shapley's systematic study of their distribution using the Cepheids as "yardsticks," convinced him that the center of distribution of the clusters was fifty thousand light-years from earth, and that this area, in the constellation of Sagittarius, was in fact the center of the Milky Way. He was right. Where Shapley went wrong was in his contention, like that of Thomas Wright almost two centuries before him, that the Milky Way comprised all there was of the visible universe—and that the apparently extragalactic nebulae were just that—apparently but not really island universes in their own right as Immanuel Kant had suggested. Shapley would cling to this belief for several years, even after Edwin Hubble had shown the true distance to the galaxy of Andromeda using Cepheids with the one-hundred-inch telescope at Mount Wilson.

Just how Shapley and Hubble and their generation of astronomers used Cepheid variables to measure the distance to such far-flung objects as globular clusters and nebulae can be traced back to the diligent work of one of the first female astronomers in the modern era, Henrietta Swann Leavitt. The daughter of a Congregational minister, Leavitt was born in 1868 in Cambridge, Massachusetts, and in 1895 would become a volunteer research assistant at the Harvard Observatory under Charles Pickering after graduating from Radcliffe. A grave illness had all but permanently destroyed her hearing shortly after she finished school. This was not considered a handicap given the job she was assigned—to study hundreds of plates of variable stars by Harvard astronomers, taken between the years 1893 and 1906 at the college's observatory in Peru. Leavitt's job was to examine and determine the magnitude, or overall brightness, of the stars. Focusing on the Smaller Magellanic Cloud, as she studied more and more plates, Leavitt noticed a relationship that applied to variable stars of the Cepheid type.

As far back as the eighteenth century, Delta Cephei in the constellation of Cepheus was seen to have a varying brightness, a very dependable varying brightness. As opposed to some binary stars that appeared to vary in brightness as one star briefly occluded the other, Cepheids were classified stars that actually seemed to swell and subside, like balloons of gas, over periods ranging from months down to merely a day. Based on careful comparison of the plates, Leavitt noticed there was a relation between the brightness of a Cepheid's magnitude and the length of time it was at its peak of luminosity: stars that were brighter apparently had longer periods of luminosity; stars that were dimmer had shorter periods. Once closer Cepheids within the Milky Way were measured by means of statistical parallax, this period-luminosity relationship allowed Shapley and later astronomers to calculate the distances to stars and nebulae farther than one hundred light-years from earth. This was astronomers' first reliable yardstick to extragalactic objects. The parallax method was suitable only for measuring distance to stars closer to Earth, those that were within a few light-years.

Leavitt died in 1921 of cancer, so she did not live to see the ultimate fruits of her momentous discovery. It was, sadly, several decades before astronomers truly appreciated the effort behind her long, patient, and very tedious work.

In the summer of 1924 when Lemaître came to Harvard, Shapley suggested that he study the theory of variable stars once he was familiar with their observations. This entailed spending much of September of the academic year at the Dominion Observatory in Ottawa, Canada. It also entailed special tutoring on the subject of Cepheid variables from a fellow countryman, François Henroteau, the former astronomer of Belgium's Royal Observatory. At Harvard Lemaître took courses in experimental spectroscopy, which relates directly to the measurement of stellar object redshifts, as well as courses on interference problems in spectroscopy.

As was noted in the last chapter, the importance of spectroscopy for astronomy and astrophysics cannot be overstated.

Lemaître, second from left, meets Hubble at Mt. Wilson in 1925; also in the photo, Hubble's wife Grace—to his left—and companion; standing at right is astronomer J. C. Duncan and presumably his wife. Man on the far left is unidentified. Courtesy of The Huntington Library, Art Collections & Botanical Gardens.

Spectroscopy allowed astronomers to determine what elements stars were made of, and it also helped them to discover the relative velocity of stars and nebulae compared to the sun. This would have momentous consequences for Lemaître and cosmology. When astronomers began to take samples of stellar spectra, they could determine what elements the star itself was (at least partially) composed of—hydrogen, carbon, or other elements. This is how helium was discovered in the sun, based on its own unique absorption lines, before it was discovered on earth. Like their counterparts in the physics laboratory, astronomers began to build up an impressive library of stellar spectra, with each star and nebula essentially having its own unique spectrum to tell astronomers what it was composed of.

After Huggins determined the relationship between the redshift of a stellar source and its apparent velocity using the Doppler effect, over the two or three decades leading up to Lemaître's work, astronomers also noticed that many of the spectra—especially those of nebulae—were shifted to the red end of the spectrum, where light consisted of longer wavelengths, eventually fading away into the invisible end of infrared and radio waves. To be sure, some stars and nebulae were shifted in the other direction, to the bluer, shorter wavelength end of the spectrum, indicating apparent velocity toward rather than away from the sun. No overall pattern was immediately recognized, except the suggestion of a radial velocity. By the time Lemaître was studying with Shapley at Harvard, astronomers were beginning to note that most of the nebulae they registered were shifted to the red end of the spectrum. Most stars and nebulae seemed to be receding, and based on their spectra, receding at truly unearthly velocities into space.

Lemaître's courses in physics did not overshadow his continuing interest in higher mathematics. He studied functional analysis and the theory of integration in classes given as a guest lecturer by his old teacher Charles de la Vallée Poussin from Belgium. As if he didn't have enough to keep him occupied, Lemaître took advantage of his year at Harvard to enroll as a PhD candidate at Harvard's rival, MIT (Harvard Observatory apparently not being able to grant him the degree).

His thesis advisor, fellow Belgian Paul Heymans, did not interest himself in questions relating to Lemaître's field, but was supportive of his program. Within the course of this year, Lemaître followed Shapley's advice and studied the theory of variable stars at Harvard. At the same time, he studied Eddington's attempt to unify general relativity with electro-magnetism in a surprising and much overlooked early attempt at unification theory. Einstein himself would later encourage Lemaître to build on this theory. And finally, Lemaître continued to work on extending general relativity to cosmological questions. As Dominique Lambert wrote, Lemaître now had all the ingredients he needed for his life's work. A visitor to Harvard Yard can still find the house where Lemaître stayed in 1924–1925. It was at 1 Cleveland Street, five blocks from the Yard's east (Quincy Street) gate. In spite of his workload, Lemaître still had to find time to fulfill his clerical obligations, which he did, often concelebrating Mass at St. Paul's Church, the parish for Catholics at Harvard.

With Shapley's encouragement, Lemaître perfected a new method for looking at a Cepheid of a given magnitude and spectral type and for calculating the period of its oscillations. His approach relied on Eddington's theory of stars which the British astronomer had applied to Cepheids in order to confirm the validity of the relationship between the mass of a star and its luminosity. Lemaître found a way to simplify the calculations of the period and at the same time to use a fitting graph for a quick reading of the result. As we noted, preliminary spectroscopic measurements of some nebulae—for example, M31, the Andromeda Nebula—made by Vesto Slipher at the Lowell Observatory seemed to indicate that quite a few of these gaseous-looking entities were receding at significant speeds away from the sun. But the measurements were at first deemed useful only to the task of helping astronomers determine the speed of the sun relative to other stars. As Slipher compiled more and more radial velocities for globular clusters and extragalactic nebula, however, astronomers realized that some other effect—subtracting the determined velocity of the sun—was at work behind the astounding radial velocities of these nebulae. And by 1922 a German astronomer, Carl Wilhelm Wirtz, was the first to rather "hesitantly" suggest that these velocities meant an actual systematic recession of nebulae away from the sun.[28]

Apparently, none of this was known in the world of cosmological theory. By 1922 Friedmann had drawn up his first model of an expanding universe with Einstein's equations. But if he was aware of Wirtz's paper, or indeed of any astronomical observations that might buttress his model of an expanding or collapsing universe, Friedmann never said so. Indeed what information we do have, according to Helge Kragh's account, is that Friedmann never considered his work in general relativity to have any physical application; to him it was only an interesting mathematical problem.[29] And it remained that way until his untimely death in 1925. But by that time, developments in astronomy were picking up momentum, and Lemaître was keeping up to date on them. Edwin Hubble had by 1924 already distinguished himself at the one-hundred-inch telescope at Mount Wilson by photographing enough nebulae to categorize them into basic categories that remain the standard to this day. And he was about to make a more striking impression on the world community of astronomers—Lemaître would be on hand to witness it.

During the summer just prior to beginning his work at Harvard, Lemaître embarked for Toronto to meet Eddington at the meeting of the British Association for the Advancement of Science. It was here that he listened to a presentation by Ludwig Silberstein, a Polish physicist studying in England, who claimed to have derived a linear relation between radial velocity and distance for extragalactic nebulae considered in de Sitter's stationary model of the universe. Even though by this time some scientists accepted the existence of galaxies as separate island universes apart from the Milky Way, it was still a point of controversy. Hubble himself never used the word "galaxy" even up to the day he died. Lemaître was inspired by this connection between radial velocity and distance. Unfortunately, Silberstein got into trouble because he claimed that his formula was confirmed by the then-current observations of globular clusters. But it soon turned out he was excluding data that did not agree with his contention. This did not escape notice. Not only were most astronomers unimpressed, but his exclusion of the disagreeable data also quite possibly prejudiced many against any argument for a relationship between radial velocity and distance of extragalactic nebulae when it was more rigorously calculated later in that decade by Lemaître and Hubble. Nevertheless, Silberstein's presentation in Toronto impressed Lemaître, who began thereafter to take a closer look at de Sitter's model of the universe using Einstein's field equations. Much later (1963) Lemaître would write in a note that Silberstein's errors had been very stimulating to him, and that he had long discussions with the Polish physicist about the issue that inspired his work as well as the work of American physicist Howard P. Robertson, who would also independently suggest dynamic models of the universe with general relativity shortly after Lemaître's 1927 paper. Within a matter of months, new data would galvanize him even more.

In early 1925, Lemaître traveled to Washington, D.C., for a lecture that would change the face of his research. Hubble forwarded a paper there showing he had resolved the existence of individual Cepheid variables in the M31, showing by that yardstick that this Andromeda Nebula was 280,000 parsecs away—much too far to be considered part of the Milky Way, and indeed much too far to be considered anything other than an island universe in its own right. In the event, his paper was read by an initially bemused Alfred North Russell, who expected the younger astronomer to deliver his paper in person and was irate when he didn't receive any text until almost the last minute. Russell was a distinguished astronomer who had codeveloped what came to be known as the Hertzsprung-Russell diagram of stellar evolution, most familiarly reproduced by a graph that plots a star's visual luminosity against its position on the sequence of evolution, from small, faint young stars through bright, massive red giants. Hubble's data now convinced most astronomers that the Milky Way galaxy was not in fact the universe in its entirety, as Harlow Shapley had once argued, but that other island universes existed independently of it, as Kant had suggested back in 1755. Andromeda, which over the ensuing years would henceforth be known as the Andromeda galaxy, was the key.

The significance of Hubble's finding was not lost on Lemaître. If other nebulae were also extragalactic, it should in theory be possible to measure their spectra to confirm if in fact they were shifted to the red end of the spectrum as others at that time seemed to be—and as Silberstein had obviously hoped. For Lemaître, the redshifts of extragalactic nebulae had to have a physical meaning, but this was not necessarily the case with his contemporary cosmologists. In his interpretation of Einstein's field equations, Willem de Sitter had posited an effect of redshift in his solutions, but he apparently thought such reddening was due to the peculiarity of the metric he derived for his cosmological solution. It did not occur to him that the expansion of space should be applied to the universe as a whole. Or, if it did, he did not take it very seriously.

Lemaître hit on this idea, but he wanted to accumulate more evidence before writing down his ideas. Never one to sit still for long, he visited Slipher in Arizona and Hubble at Mount Wilson in the summer of 1925. Hubble's wife, Grace, was a tireless journalist in the original sense of the word—she kept detailed day-to-day accounts of her husband's life and career up until the day he died.[30] Although she has an opinion about almost everyone they met, and we can see her and Hubble along with Lemaître in a photo taken at the California Institute of Technology (better known as Cal Tech) at the time, we do not know Hubble's personal impressions of Lemaître—or his wife's. This is unfortunate, considering the amusing and often gossipy nature of her impressions of so many other of the personalities she and her husband encountered. As Kragh writes, Hubble was a cautious scientist, and even when it became clear to most astronomers that his discovery of the recession of the galaxies clearly supported Lemaître's groundbreaking paper on the expansion of the universe, Hubble rarely mentioned Lemaître by name in his own articles and stubbornly resisted any outward direct support for expansion models of the universe. In hindsight this seems not just overly cautious but downright timid. And if his wife's notes are to be taken at face value, it's clear that Hubble had a low opinion of theory in general.[31]

However the visit went in 1925, we do know that Lemaître took away some fresh data on redshift observations and that by the time he was at home again in Belgium at the start of the 1925–1926 academic year, he was ready to submit his first paper attacking the unsatisfying aspects of Einstein and de Sitter's "static" models of the universe.

5.

EXPANSION IS DISCOVERED

AS MANY ACCOUNTS point out, when Arthur Stanley Eddington and the Royal Astronomical Society announced confirmation of the bending of starlight by the sun, a phenomenon predicted by the general theory of relativity and laboriously photographed during the 1919 solar eclipse on the island of Principe, Albert Einstein became an overnight sensation. For the rest of his life Einstein would be a figure of legendary intellect—and controversy. Unfortunately, because of the publicity at this time, some aspects of his work would also acquire a lifelong—indeed, century's long—aura of almost mystical difficulty. One reason for this mystique was the media's often unquestioning infatuation with Einstein.

And coverage from the start was bungled, with the decision by the *New York Times* to send their London office's golf correspondent to cover the story of Einstein's new theory, since no one with any scientific credentials could be sent on short notice. As David Bodanis writes in his book $E = mc^2$:

> The *New York Times* did have a few knowledgeable science writers, but they were in New York. The London bureau was handed the story, and Henry Crouch was asked to cover [it] . . . Crouch was a good journalist in the sense that he knew that you had to make a story interesting. He was somewhat less good, however, in having the slightest clue what was going on here— Crouch was the paper's golfing specialist.
>
> But he was also a *Times* man through and through, and nothing like a simple lack of knowledge was going to hold him back. He kept on filing, and the headline writers pulled out the key parts of his story:
>
> "A Book for 12 Wise Men: No More in All the World Could Comprehend It, said Einstein When His Daring Publishers Accepted it."[32]

Einstein was not writing a book about relativity—scientific theories were not proposed that way, as any scientist could have explained to the *Times* correspondent. Papers were published in the reputable journals of the day, such as the German *Zeitschrift für Phyzik,* or the British *Monthly Notices of the Royal*

Astronomical Society. Other scientists then followed up with criticism, or figured out ways to test a new theory and then published their own results (which is, essentially, what Eddington did in Einstein's case). But there were no "daring" book publishers involved here. Crouch simply invented the story for the sake of dramatic effect. The golf reporter found the Swiss physicist baffling and contented himself with puff pieces breathlessly claiming that only a handful of people in the world could understand this strange new theory of the universe. (Eddington deserves some of the blame, too, for doing his part to play up the mystique.[33]) It would not be an exaggeration to say that the popular consciousness of relativity's vaunted difficulty has not recovered from this hyped-up journalism.

Ever since, Einstein's field equations for general relativity have taken on a frightening aura of almost mystical impenetrability. But in fact their degree of difficulty has more to do with the level of application—the time and patience required to solve many differential equations in the geometry of space-time—than with the conceptual difficulty of what Einstein was trying to do. This is not to say that anyone with a year's level of college calculus can tackle the field equations, but there is a standard path of progress for any student to take—multivariable calculus, differential geometry, and the lingo to go with it—in order to become equipped to study general relativity. And thousands of students have done exactly that.

In classical Newtonian theory, an object's path in the presence of gravity is determined by its mass (in kilograms, for example) and its acceleration. Mass is also key to determining how objects travel in Einstein's theory, but his equations do not specifically treat an object's mass alone. Rather they determine how the curvature of space-time, which determines an object's path of motion, is influenced by the object's mass *and* its energy *and* its density. At first this may seem strange, since it's difficult to imagine features like pressure and energy having mass in concrete terms, but it's important to remember from the theory of special relativity that the famous mass-energy relation, $E = mc^2$, demands this be the case: even energy can have weight, in a manner of speaking.

In general relativity, energy and density are treated on the right side of Einstein's field equations, while the left side of the equations treat the curvature of space-time that results from it. The cryptic equation $R_{\mu v} - 1/2g_{\mu v}R = -8\pi GT_{\mu v}$ boils general relativity down to this basic relationship, a summation of what general relativity means. And certain classic symbols represent the quantities to be derived. On the left side of the field equations of general relativity, $g_{\mu v}$ represents the Riemann tensor for the measured distance between two points in curved space, tensors in essence here being groups of vectors. $R_{\mu v}$ represents the Ricci tensor, which determines space-time curvature. The two tensors are named after the mathematicians who invented them, Bernhard Riemann (1826–1866) and Gregorio Ricci (1853–1925). Both terms are summed over the subscripted indices, represented by the Greek letters μ (mu) and v (nu), which stand for variables that change over the four coordinates representing the dimensions of space—*x, y, z*—and time—*t*—at each point. R stands for the trace of the Ricci tensor.

Lemaître, circa 1930. Photo courtesy of Archives Lemaître, Institut d'Astronomie et de Geophysique Georges Lemaître, Catholic University Louvain.

On the right side of the equations, G represents Newton's constant of gravitation, T represents the stress-energy tensor, also summed over the indices, which reveals the properties of matter, energy, and momentum over the distance of curved space in question. In cosmological applications of the field equations, one also finds r standing for the radius (of a sphere or cosmological model, such as discussed by Schwarzschild, Lemaître, and de Sitter), t the time, p the pressure, and the Greek letter ρ (rho), for density. There are many other symbols used in Einstein's equations, but these were the most common in the papers on relativistic cosmology of the period in question.

It's important to note that the practical solutions to the Einstein equation, commonly called "metrics," for the measure of distance between points in curved space-time, are more recognizable to math students. Lemaître's line element, for example, from which he derived the solutions in his 1927 paper, is the quadratic equation $ds^2 = -R^2 d\sigma^2 + dt^2$, where, as he wrote, "$d\sigma$ is the elementary distance in a space of radius unity, and R [radius] is a function of the time t."[34] With Einstein's tools in hand, a physicist can determine how the presence of mass, energy density, time, and pressure influenced the geometry of space-time. He or she can not only determine the path of objects in a gravitational field, but can also apply them consistently to the universe as a whole. This had not been the case with Newtonian theory. Indeed, Newtonian gravitation's inconsistencies, when applied to the universe on the grand scale, discouraged the theoretical exploration of any consistent cosmology until the time of Einstein.

As we have seen, Lemaître, who started out as a student of mathematics, quickly absorbed Einstein's equations during the years 1920–1923 while he was studying for his degree and preparing for the priesthood. But it's clear from his papers and correspondence with his mentors that he also had a facility for finding easier ways to compute and determine solutions to the equations than many of his peers. By 1925, when he had visited Slipher at Flagstaff and Hubble at Mount Wilson to discuss their observations, Lemaître understood the tools and the implications of general relativity so well, he was ready to apply the field equations to his own "cosmological considerations." He did this at first, mainly by pointing out "problems" with the solutions that had been published before, meaning specifically the cosmological solutions to the equations published by Einstein and de Sitter in 1917 and shortly after.

The Lemaître/Eddington model circa 1931

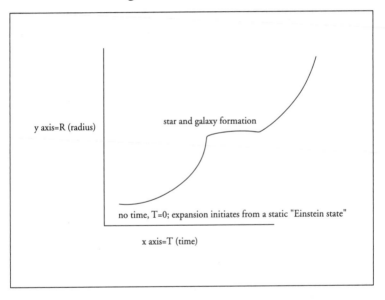

y axis=R (radius)

star and galaxy formation

no time, T=0; expansion initiates from a static "Einstein state"

x axis=T (time)

Rough illustration of the Lemaître/Eddington model-drawing by the author: this model starts from a static Einstein state which extends into the infinite past and rapidly expands; it undergoes a pause, or gestation period during-which stars and galaxies form, before expansion resumes at an accelerated pace into the de Sitter universe of flat space.

This was Lemaître's style, following a pattern in his life that we witnessed previously. You may recall from chapter 2 how during his artillery training Lemaître angered the ballistics officer by pointing out what he considered an obvious mistake in one of the equations in the ballistics manual. For his insolence, Lemaître and his brother were thrown out of class for insubordination and issued a reprimand that undermined Lemaître's chances for making officer grade. Thus it's tempting to see Lemaître's first cosmological paper similarly pointing out "errors" in Willem de Sitter's solution to the Einstein field equations. Strictly speaking, there weren't errors, but de Sitter drew conclusions about his model of empty space drawn from a mathematically limited point of reference that Lemaître suggested could be enlarged upon with clarifying results. In a way, Lemaître was doing what he had always done—noticing mathematical inconsistencies and flaws and seeing how their correction opened up new ways to look at the problem. (This may be why at least one historian referred to Lemaître as "plump, irritating, and ahead of his time."[35]) This is what he did in 1925 with his first key paper, "Note on de Sitter's Universe."[36]

As we saw in the first chapter, in 1917, within just a few months of Einstein's paper on cosmological considerations, Willem de Sitter submitted a paper to the *Monthly Notices of the Royal Astronomical Society* titled "On Einstein's theory of gravitation and its astronomical consequences," arguing that one could use Einstein's equations to posit a **static** universe that was completely empty of matter (specifically devoid of energy density).[37] De Sitter's paper essentially set the terms of the first debate in early relativistic cosmology for a decade and more. The debate was seen essentially as a choice for one of two possible universe models:

- Einstein's static spherical world, the finite but unbounded sphere of limited matter, which Einstein believed to be the *only* solution to his equations.

- De Sitter's expanse devoid of matter and energy. (As was mentioned in chapter 1, no one took any notice of Alexander Friedmann's solutions of 1922 and 1924 that described both expanding and collapsing models of the universe. And Einstein did his best to forget about them.)

Einstein fundamentally disagreed with any solution to his equations that suggested a model of empty space. Space devoid of matter not only made a hash of his assertion that the curvature of space-time was determined by matter and energy, but it therefore also contradicted the assertion that matter's path in a gravitational field was determined by space-time curvature. Needless to say, de Sitter's solution also undermined Einstein's contention that his static universe model was the *only* cosmological solution possible to derive from his field equations. On that score, Einstein was going to be continually surprised for the next two decades, as more and more physicists came up with their own solutions and models from his equations.

For the time being, Einstein argued, how could it be possible to have a universe of space-time with no matter whatsoever to occupy it? That seemed to be the debate for the period of 1917 through the end of the 1920s, and physicists and mathematicians who entered the fray with their own papers, such as Eddington in the United Kingdom, Hermann Weyl in Germany, and Richard C. Tolman and Howard Robertson in the United States, all put forth their arguments based on the question of whether Einstein's or de Sitter's solution was the more likely. Aside from poor Friedmann, who died in 1925, and Lemaître, who for most of this time was ruminating and not writing on the issue, no one seems to have considered any other nonstatic model. And this state of affairs persisted right up until the dawn of the New Year in 1930.[38]

In hindsight this may seem odd, but it's important to bear in mind two points about the period. One is that cosmology in a sense didn't even exist as a discipline. There were no programs of study in cosmology. There certainly weren't any journals dedicated to cosmology. Einstein's 1917 paper launched the discipline as we now know it, but for the next few decades the number of specialists contributing to the discussion was very small compared to the number of specialists working, for example, on the exciting new quantum theory of atomic physics. Interest in exploring atomic structure and building the theoretical system of quantum mechanics attracted much more interest among students and professors than general relativity, to say nothing of cosmology. It also attracted more research grants. Second, as previously noted, astronomical evidence to support Einstein's general theory of relativity basically boiled down to two phenomena: the bending of starlight and the advance of the perihelion of Mercury. As much energy and ink (if not more) was spent by astronomers and physicists trying to explain these matters by modifying the older, classical Newtonian theory as was spent trying to figure out the cosmological consequences of Einstein's new theory. Thus between 1917 and 1927, Einstein and de Sitter cordially swapped papers over this "controversy," defining the central question of cosmology at the time: not whether the universe was static—both scientists and most of their colleagues continued to assume it was—but which universe, Einstein's or de Sitter's, made more sense based on the new evidence that was slowly accumulating from the American telescopes in California and Arizona.

It was here Lemaître first entered the fray with his 1925 paper (it was also a part of his PhD thesis submitted to MIT). De Sitter's solution had certain known side effects, which in hindsight were clear indications that any truly useful model of the cosmos in general relativity had to be dynamic, not static. Specifically, de Sitter's solution showed that any particle introduced into his empty static universe would appear to recede from any other particle and show some redshift. This was a tantalizing premonition of the redshift that would be found to represent the actual recession of the galaxies discovered and reported by Hubble in 1929, but it was not recognized as such at the time. Eddington, for example, considered this phenomenon, which he called the **de Sitter**

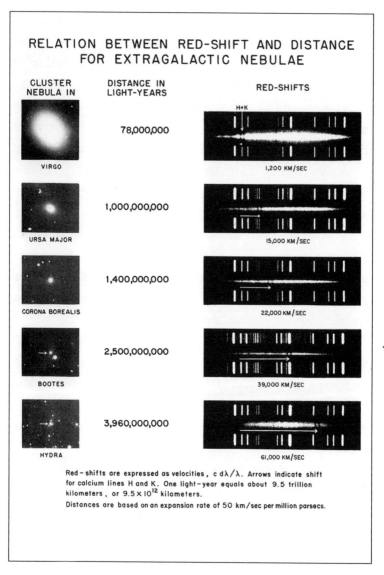

The redshift velocity distance relation. Courtesy of the Observatories of the
Carnegie Institution of Washington.

effect, as only apparent. He suggested that the ostensible stretching, or reddening of light from particles introduced in the de Sitter universe, was caused by a metric effect due to the curved spacetime geometry. In his 1925 paper, Lemaître, who was already thinking about an expanding universe model, noticed that one of the reasons de Sitter argued that his model was a "static" solution was that he had made a mistake in his equations, picking a preferred frame of reference for his argument. In effect, de Sitter was assuming a lack of homogeneity in a point of space, one that theoretically should not have been there. Both the Einstein universe and the de Sitter universe were supposed to be homogeneous and isotropic for theoretical purposes, which is to say they were supposed to be free of privileged points and free of privileged directions according to the principles of the theory of relativity. Lemaître showed that de Sitter had, in effect, accidentally adopted a privileged position in his space, a reference frame that was "special," and Lemaître argued that this is why the Dutch astronomer had concluded, wrongly, that such a universe would feature, for example, an impermeable horizon, beyond which one could not see and at which, as Eddington pointed out, all the processes of nature would come to an apparent standstill. Lemaître showed that by changing coordinates one could preserve the **homogeneity** and **isotropy** of the universe under general relativity—and also showed that the scale factor, or radius, of this universe, *would not be constant,* as was the case in both Einstein's and de Sitter's original solutions. Indeed, Lemaître showed that the radius would be a time-increasing function, that distances between all points in space would increase.

So de Sitter's universe would be empty—but expanding. And more to this, Lemaître showed that while space was homogeneous, it was no longer closed as in Einstein's spherical model. It would be an infinitely extending, spatially flat, space. In short, Lemaître showed that de Sitter's universe was in fact a limited case of an expanding universe model. And indeed, although at the time de Sitter considered his model to be static, it is no longer considered such by historians of science, since Lemaître's "Note" showed that the "static" nature was in essence a mathematical fiction. And Lemaître's interpretation of de Sitter's universe became a key foundation for the later **steady state theory** and for the inflation theory.

Lemaître had now written a sort of prologue to his own revolutionary paper two years later that, like Friedmann's before him, would offer a completely new dynamic model for the universe. But there was one essential difference: **Lemaître's model** entailed an evolving universe, with red-shifted nebulae illustrating space-time expansion, building from both Einstein's and de Sitter's models, and expanding according to a law of nebulae receding at radial velocities directly proportional to their distances—a law that would eventually bear Edwin Hubble's name.

It's important to note once again the key difference between de Sitter's model and what would become Lemaître's dynamic model. De Sitter's space did indeed suggest a recession of objects away from each other, and the Dutch astronomer pointed out that some of the redshifted spectra assembled by astronomers already at that time, in the early 1920s, could be interpreted as due to the de Sitter effect. But de Sitter did not, nor did anyone else, attribute the recession *to the actual expansion of space-time* itself. De Sitter's work was pointing to something more momentous than he realized, according to Lemaître.

What immediate effect did Lemaître's 1925 paper have on the astronomical community at the time? Apparently little. What we know of de Sitter's response, from Andre Deprit's biographical article on Lemaître, has more to do with Lemaître's paper of 1927, and at that point apparently de Sitter had no time for Lemaître's work when the latter approached him during the Third General Assembly of the International Astronomical Union in Leiden, which took place from July 5 through July 18, 1928. De Sitter had by then become president of the union and was, according to Deprit, feeling fairly full of himself (although no documentation exists to support this admittedly personal impression). As for Einstein, he was already wrapped up in exploring preliminary versions of a unified field theory that he would spend the rest of his life trying to formulate. He was also becoming increasingly politically active, at least in terms of his pronouncements to the mobs of reporters who waited for him at every stop of his increasing travels. Aside from his correspondence with de Sitter, Einstein took little notice of what his theory was inspiring among other physicists and astronomers.

One can almost feel Einstein's impatience with the whole business of general relativity's cosmological consequences during this period. The almost terse half-page criticism he submitted (and then retracted) to *Zeitschrift für Phyzik* regarding Friedmann's expanding model solutions in 1922 and 1924 are an example. Einstein clearly felt he had more pressing questions to attend to, like Bohr's interpretation of quantum mechanics. That would change once Lemaître tracked him down at the Solvay conference in October 1927.

Once Lemaître finished his "Note on de Sitter's Universe," he settled down—however briefly—at his new teaching post as professor of physics at the Catholic University of Louvain, where between 1925 and 1927 he worked out its sequel—the details of a complete solution to Einstein's equations that would fully model an expanding universe. The paper had the lengthy title of "A Homogeneous Universe of Constant Mass and Increasing Radius Accounting for the Radial Velocity of Extra-Galactic Nebulae," and began by recapping some of what Lemaître had found wanting about de Sitter's universe:

> When we use co-ordinates and a corresponding partition of space and time of such a kind as to preserve the homogeneity of the universe, the field is found to be no longer static; the universe becomes of the same form as that of Einstein, with a radius no longer constant but varying with the time according to a particular law.
>
> In order to find a solution combining the advantages of those of Einstein and de Sitter, we are led to

consider an Einstein universe where the radius of space or of the universe is allowed to vary in an arbitrary way.[39]

Lemaître wanted, from the outset, a cosmological solution that would nicely enfold both Einstein's and de Sitter's solutions as limiting cases of a larger model. Such a universe model needed to be closed—meaning finite but unbounded according to Einstein's initial solution—as well as homogeneous and isotropic, and with a positive curvature. Most importantly, however, Lemaître also wanted an expanding model of the universe that would accommodate the data from

Lemaître with Einstein and Robert Millikan at Pasadena, 1933. Photo courtesy of Archives Lemaître, Institut d'Astronomie et de Geophysique Georges Lemaître, Catholic University Louvain.

the existing astronomical observations of redshifted nebulae he had gathered from his talks with Hubble and Slipher.

This is not a trivial point, for in the early years of general relativity, the 1920s and 1930s, the chasm between the mathematical abstraction of the theory and physical observations and tests of such was substantial. For many physicists and astronomers, the cosmological models derived from the Einstein field equations were considered purely speculative. And with the growing fascination and interest in quantum mechanics and nuclear fission, general relativity and cosmology would remain for these decades the rarefied interest of a comparatively few physicists.

Rarefied as it seemed, Lemaître rooted his model as much as he could in the data available to him. In fact, using Hubble's estimates of the time, he obtained a radius for his model of the universe at $R_E = 8.5 \times 10^{28}$ cm, which equals 2.7×10^{10} parsecs.

It's important to note here that Lemaître was not yet interested in discussing a temporal beginning of the cosmos in any sense of the term, as has often been inaccurately stated in many books about twentieth-century cosmology. His 1927 paper suggested an expansion of the universe beginning from an initial static Einstein state—not a big bang, not an explosion of matter from nothing. (Indeed Lemaître, having some philosophical training, never made the mistake of pretending that "creation" could be defined in scientifically meaningful terms.) He was interested in showing how the current universe could be a model that started from Einstein's static solution, a model that could have existed for an infinite time into the past, and then evolved, expanding ultimately into the flat **de Sitter model.** His further theory of a **primeval atom,** or initial cosmic origin, would come later, once he realized the physical deficiencies of beginning with Einstein's model.

Lemaître with Einstein at Pasadena, perhaps discussing the cosmological constant. Photo courtesy of Archives Lemaître, Institut d'Astronomie et de Géophysique Georges Lemaître, Catholic University Louvain.

When Lemaître finished his paper, he made the curious decision to submit it to a relatively obscure Belgian journal, the *Annales de Societé Scientifique de Bruxelles,* rather than one of the more widely read journals, such as *Zeitschrift für Phyzik.* Neither, apparently, did he consider sending it to Eddington for possible publication in the *Monthly Notices of the Royal Society in England* (as de Sitter had done with some of his own papers). Unfortunately, the venue Lemaître chose for his groundbreaking paper would almost guarantee that no one would take any notice of it, and this fact has puzzled historians. (On the other hand, given the utter lack of interest in Friedmann's papers, which were published in one of the most widely read physics journals of the day—published and essentially ignored even when Einstein deigned to respond to them—perhaps it should not be puzzling at all.) It has been suggested that Lemaître published in the smaller circulation Belgian journal because he was somewhat hesitant to draw attention to his model, knowing just how speculative and provocative it was; that he didn't really want to draw attention to the heterodox cosmological speculations he was entertaining. And there could be some truth to this. From the vantage point of today, where the **big bang** model is considered standard and the expansion of the universe is taken for granted, it's difficult to appreciate just how radical Lemaître's model would be considered back in the 1920s, especially before Edwin Hubble published his findings that the majority of galaxies measured were receding from the sun.

Einstein's generation of physicists—including Max

Planck—were of the nineteenth century in the sense that they had grown up with the normalcy of a static universe they inherited from centuries of European thought. And it did not occur to them to change it. It did not occur to Einstein, whose own field equations first suggested that any model of the cosmos would be dynamic. Still, such a suggested hesitancy on Lemaître's part doesn't quite jibe with what we know of his intense interest and activism of the years immediately prior to his paper—the traveling all over the United States and Canada, for example, to talk with all the leading astronomers, such as Hubble, Slipher, and Shapley; the influence of Silberstein's misfired paper; the constant discussions with Eddington, who was always interested in his pupil's work. These adventures do not sound like the efforts of a man who was hesitant to make his own contribution to the field. It may be that Lemaître sent his paper to the Belgian journal in something of a rush because he knew it would be published sooner than later. And for that reason, he may have regretted his choice in light of what followed. We just don't know. What we do know is that what followed over the next two years was silence. Even though Lemaître sent a copy right away to Eddington, his former mentor either put it aside for later reading and then forgot about it, or more likely read it and did not then appreciate its revolutionary nature. In any case, Lemaître's paper would languish for more than two years.

During this time, Richard C. Tolman, a physicist at Cal Tech whom Lemaître had met in his early trips to Pasadena,

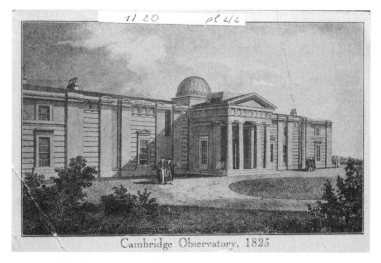

Cambridge Observatory, 1825

Front face of Eddington postcard to Lemaître, 1933. Image courtesy of Archives Lemaître, Institut d'Astronomie et de Geophysique Georges Lemaître, Catholic University Louvain.

and with whom he would later collaborate, made his own contribution to the question of the Einstein and de Sitter solutions in 1929. Tolman rediscovered what Friedmann had shown in 1922—that Einstein's and de Sitter's solutions were the only nonstatic solutions to the field equations. A year before, another American physicist, Howard Robertson, unaware of Lemaître's "Note" of 1925, independently reached the same criticism of de Sitter's solution as Lemaître had.

The next step awaited the results of Edwin Hubble's monumental work. Already in 1925, Hubble had astounded the world of astronomy with his determination, based on Cepheid variables he had located in M31, the Andromeda Nebula, that this nebula was on the order of eight hundred

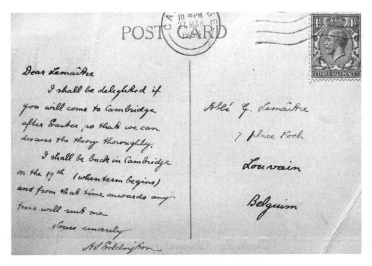

Dear Lemaître

I shall be delighted if you will come to Cambridge after Easter, so that we can discuss the theory thoroughly.

I shall be back in Cambridge on the 19th (when term begins) and from that time onwards any time will suit me

Yours sincerely

A.S. Eddington

Abbé G. Lemaître

7 Place Foch

Louvain

Belgium

Back side of Eddington postcard, 1933, discussing a prospective meeting to discuss one of Lemaître's theories. Photo courtesy of Archives Lemaître, Institut d'Astronomie et de Geophysique Georges Lemaître, Catholic University Louvain.

thousand light-years away—showing that the universe was indeed much larger than just the Milky Way, and that our home galaxy was just one galaxy among many.

Now Hubble pressed on with the detailed study of more extragalactic nebulae and their redshifts. A pattern was already suggested, as we saw, by the work of Vesto Slipher before 1920, showing a handful of nebulae with redshifts indicating recessional velocities away from the sun of awe-inspiring speeds. But many more would need to be expertly catalogued, and those already taken carefully reexamined, in order for Hubble to make any connection to either Einstein's or de Sitter's model of the cosmos.

Between 1925 and 1929, Hubble and his dogged assistant,

Milton Humason, spent month after month using the one-hundred-inch reflector at Mount Wilson to painstakingly accumulate new redshifts from as many nebulae as they could. In the end, they published their results using twenty-five spectra out of forty-six taken for galaxies. Hubble published his famous paper in 1929, modestly titled "A Relation between Distance and Radial Velocity among Extra-Galactic Nebulae."[40] In a model of understatement, he submitted that:

> The results establish a roughly linear relation between the velocities and distances among nebulae for which velocities have been previously published, and the relation appears to dominate the distribution of velocities . . .
>
> The outstanding feature, however, is the possibility that the velocity-distance relation may represent the de Sitter effect, and hence that numerical data may be introduced into discussions of the general curvature of space.

The farther away a galaxy was measured in space, the faster it appeared to be receding. Hubble submitted a mean velocity of recession at 500 km/sec. This is not as fast as the 625 km/sec figure Lemaître derived in his 1927 paper two years before, but it's not too far from it.

The gauntlet was now in effect thrown back down at the feet of the theorists. It became clear that the relation between distance and apparent velocity did not fit either Einstein's solution or de Sitter's. No one knew what to do until Eddington publicly mulled over the need to reconcile the Einstein and de Sitter solutions in some new way that accommodated **Hubble's law.** This he did in a meeting of the Royal Society in January of 1930. (He is said to have quipped at the time, "Well, shall we put a little motion into Einstein's universe or shall we put a little matter into de Sitter's?") The notes from the proceedings were published a few months later. When Lemaître read them in Belgium, he immediately sent his former mentor a message (along with another copy of his 1927 paper), telling him that he had in fact already solved this problem.

A pupil of Eddington's at the time, George McVittie, recalled the scene many years later when he was asked to write Lemaître's obituary notice for the Royal Astronomical Society:

> I well remember the day when Eddington, rather shamefacedly, showed me a letter from Lemaître which reminded Eddington of the solution to the problem which Lemaître had already given. Eddington confessed that, though he had seen Lemaître's paper in 1927, he had completely forgotten about it until that moment. The oversight was quickly remedied by Eddington's letter to *Nature* of 1930 June 7, in which he drew attention to Lemaître's brilliant work of three years before.[41]

With the new solution in hand, Eddington also had Lemaître's paper translated and published in the *Notes of the Royal Astronomical Society.* De Sitter, slow to respond just a few years before,

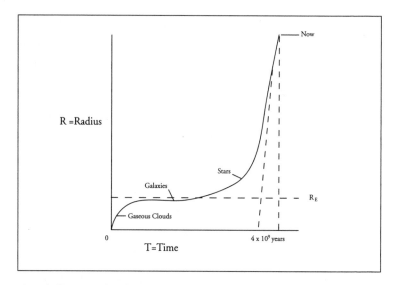

Rough illustration based on graph from Godart and Heller's 1985 book, showing Lemaître's model from the mid 1930s—starting from Time $t = 0$ and expanding with a positive cosmological constant into the present period. According to Godart, the current radius of the universe (meaning at the time Lemaître devised this) is assumed to be about ten times that of the radius of the Einstein universe (R_E). The timescale of 4×10^9 years, is derived from the value of the Hubble constant as it was estimated at that time.

now immediately grasped its importance, as did everyone else: Lemaître had at a single stroke created the first consistent, evolving model of the universe that beautifully incorporated Hubble's findings on the apparent velocity-distance relation (even before Hubble had published them). Hubble's law might just as easily have been called Lemaître's law.

Lemaître's public life was about to begin. His solution seemed made to order, and the avowed expansion of the universe was no longer a mathematical contrivance. It was a reality.

6.

THE PRIMEVAL ATOM

As a scientist, I simply do not believe the universe began with a bang.

Arthur Stanley Eddington, The Nature of the Physical World

A complete revision of our cosmological hypothesis is necessary, the primary condition being the test of rapidity. We want a "fireworks" theory of evolution. The last two thousand million years are slow evolution: they are ashes and smoke of bright but very rapid fireworks.

Georges Lemaître, "The Evolution of the Universe" [5]

IN HIS LATER career, Lemaître often had to deal with the suspicions of scientists—astronomers and physicists such as

Fred Hoyle and William Bonnor—who suspected that his primeval atom theory was inspired by religious faith: that his training as a Catholic priest somehow biased him toward speculating about the cosmic origin of the universe in a cosmic fireball evocative of the creation story from the Book of Genesis.

At the Athenaeum at Cal Tech in Pasadena in January 1933, Lemaître noticed that Einstein was much more impressed with his work than he had been at their previous meeting in Leopold Park in 1927 during the Solvay conference. Lemaître had come to give two seminars—one a review of his 1927 paper on the expanding universe as a solution to the problem of the Einstein and de Sitter universes, the other, more important to the Belgian physicist, on cosmic rays and their potential role as fossils left over from the super-dense cosmic quantum at the origin of the universe.

Much later, in 1958, in a short radio address recounting his friendship with Einstein, Lemaître recalled how Einstein had complained in one of their meetings that Lemaître's expanding cosmic nucleus was unacceptable because of its metaphysical implications:

> As I spoke with him about my ideas regarding the origin of cosmic rays, he said excitedly, "Have you spoken with Millikan?" but when I spoke to him about the Primeval Atom, he interrupted me, "No, not that, that suggests too much the creation."[41]

Robert Millikan, who taught at Cal Tech, was at the time the foremost expert studying cosmic rays, and he and Lemaître would correspond about them over the ensuing years. Lemaître believed cosmic rays might be the leftover "fires and smoke" he had mentioned of the super-dense cosmic nucleus from which he theorized the entire universe evolved. Clearly Einstein must have understood their importance to his theory, so his sudden outburst against the primeval atom theory seems inconsistent.

In fact there is reason to believe that Einstein not only had no problem with the idea of a cosmic beginning, but also that he and Lemaître discussed whether it was possible to avoid cosmic singularities at the proposed origin of space and time. **Singularities** in Einstein's field equations are essentially elements, such as density or pressure, that cause the equations of general relativity to break down, due to one or more factors rising to ∞ (infinity) and in effect making the equations unworkable. He suggested to Lemaître to look into the possibility of anisotropic, nonhomogeneous models as a means of avoiding any singularity at the cosmic origin. Lemaître did indeed write a paper on this, first published in 1932 (and reprinted in 1933), but he found that the singularity was apparently unavoidable. Neither he nor Einstein, however, accepted that his result necessarily implied a physical reality to the singularity.

Later work by Roger Penrose and Stephen Hawking would prove that singularities are unavoidable in all relativistic world models. But regarding a temporal origin to cosmological models, Einstein later wrote in his *Meaning of Relativity* that "one may not conclude that 'the beginning of the expansion' must mean a singularity in the mathematical sense. . . . This consideration does, however, not alter the fact that 'the beginning of the world' really constitutes a beginning, from the point of view of the development of the new existing stars and systems of stars, at which those stars and systems of stars did not yet exist as individual entities."[43] It appears, then, that Einstein was willing to entertain the possibility that the universe may have had a temporal beginning. And, notwithstanding his quip to Lemaître, he had enough philosophical grounding to realize that an origin of space-time was not the same thing as creation of the world out of nothing, a concept he appreciated was intrinsically outside scientific bounds.

Now as they took frequent walks about the grounds of the Athenaeum at Cal Tech, Lemaître found Einstein willing to listen to his arguments in favor of keeping Λ the cosmological constant, in the field equations of general relativity, even though Einstein believed that Hubble's discovery of the recession of the galaxies made it superfluous now to expanding cosmological models based on his theory. Lemaître was amused to see reporters hovering not far off wherever Einstein went, watching him with the celebrated physicist as they had their discussions. It soon became a running joke that whenever Einstein and the priest went for a walk, they were sure to be discussing "little lamb" the reporters' shorthand for Λ, lambda. At the time, Lemaître smoked cigarettes. Einstein smoked his favorite pipe. His wife, Elsa, often rationed him to a limit of tobacco, however, and often Lemaître let Einstein peel apart one of his cigarettes for the extra tobacco when he ran out of his own stock.

Some accounts of Lemaître's work have perpetuated the notion that the primeval atom theory dates from 1927—and that his paper on an expanding universe model, incorporating both Einstein's and de Sitter's solutions, necessarily implied a cosmic origin from the start.[44] As we've seen, Lemaître's 1927 paper deals only with a model of a universe that expands from the already existing static Einstein model—evolving over time into the de Sitter state of virtually empty space. It was the whole idea of a dynamic universe—not its origin—that Lemaître introduced at the time, a dynamics that Einstein objected to, just as he had objected to Alexander Friedmann's dynamic models back in 1922. He said nothing to Lemaître in 1927 about any problems associated with a presumed hint of creation. That was to come later.

The origin of what has become the big bang theory is more prosaic but hardly surprising if one follows the logic of Lemaître's papers and of the field equations of general relativity. By 1930, when Eddington published the translation of Lemaître's 1927 paper, Lemaître was already having misgivings about the stability and endurance of a cosmic model that started from a static Einstein state stretching back into an indefinite past. The reason was physical: the static Einstein universe could not sustain itself indefinitely. Lemaître saw that the expansion of the Einstein universe slowed down logarithmically in the past. This implied that the very physical processes of such a universe also slowed, at the same rate. Such a universe could not—in reality—be temporally infinite. Close examination of the static Einstein model revealed there *had to be* some kind of beginning of all physical processes in order to work as a cosmological model.

Paul Dirac, a leading quantum physicist and Nobel laureate, who essentially created the theory of quantum electrodynamics the same year Lemaître wrote his key paper on the expanding model, took a more than passing interest himself in relativistic cosmology. He was well aware of the problem Lemaître encountered with his initial 1927 model, considering Hubble's data:

> [O]ne would have the expansion starting at a time of the order 10^9 years ago, which is the time provided by **Hubble's constant** [bold added] if the expansion were approximately uniform. Now one of the big difficulties then confronting cosmologists was that stellar evolution required a much longer time than this. Lemaître therefore favored his [original] model, which pushed back the start of the expansion into the infinite past.
>
> He subsequently modified this view, stating that it does not really help with the discrepancy to push back the start of the universe in this way, because in the early stages of the expansion all physical processes would take place extremely slowly, the rate slowing down log-

arithmically as one goes back to the infinite past. Thus the extra time in the early stages would not be available for much stellar evolution to occur. This kind of argument showed up Lemaître's appreciation of the need for understanding the physical significance of one's equations.[45]

In other words, there was really no way to "buy time" with a proposed static state of infinite duration on which to base an expanding model. In effect, the very evolution of the universe could not be separated from its expansion. They had to go hand in hand. So some kind of origin had to be entertained. In fact, the issue of a concrete origin began bothering Lemaître over the period immediately after he wrote his 1927 paper. By 1930 he was unsatisfied with his original model, even as his colleagues and the press publicized and celebrated it. Lemaître realized he needed to come up with a cosmological model that was more physically satisfying, but he didn't get a start on the issue until 1931.

Eddington, again, was Lemaître's inspiration for his primeval atom hypothesis. In January of that year the English astronomer gave a talk to the British Mathematical Association (of which he was the president) called "On the End of the World from the Standpoint of Mathematical Physics" in which he argued in favor of the idea of a slow heat death of the universe.[46] But he also extrapolated backward from the present state of the universe, and wondered what a "beginning" might be like:

Following time backwards, we find more and more organization of the world. If we are not stopped earlier, we must come to the time when the matter and energy of the world had the maximum possible organization. To go back further is impossible. We have come to an abrupt end of space-time—only we generally call it the "beginning" . . . philosophically, the notion of a beginning of the present order of Nature is repugnant to me.

Lemaître was struck by this last statement. In response, he submitted a letter to *Nature,* published a few months later, "On the Beginning of the World from the Point of View of Quantum Theory."[47] This was the groundwork, the first declared proposal, for what would become the big bang theory. He first introduced his theory as a "fireworks" origin of the universe. What's remarkable about this *Nature* letter is that—apart from discussing the idea of a temporal beginning of the cosmos—it marks the first time that a physicist directly tied the notion of the origin of the cosmos to quantum processes.

Lemaître had long been aware of the meagerness of models of the universe that were purely geometric, as the models based on Einstein's field equations were. They were large-scale models of geometry, and from a theoretical perspective they were satisfying. But for a mathematician who was also a physicist, they were problematic. Lemaître's colleague Tolman felt the same way, and in fact this is what restrained Tolman from more boldly attaching his own expanding model, published in 1929, to Hubble's redshift measurements the way Lemaître did.

Even at this early stage in the development of quantum mechanics, when many of the particles we now take for granted, like the neutron, had not even been discovered yet, Lemaître realized that any realistic model of the universe's origin had to be rooted at the microscopic level in quantum physics. In his May 9, 1931, letter to *Nature* he began:

> Sir Arthur Eddington states that, philosophically, the notion of a beginning of the present order of Nature is repugnant to him. I would rather be inclined to think that the present state of quantum theory suggests a beginning of the world very different from the present order of Nature.
>
> Thermodynamical principles from the point of view of quantum theory may be stated as follows: (1) Energy of constant total amount is distributed in discrete quanta. (2) The number of discrete quanta is ever increasing. If we go back in the course of time we must find fewer and fewer quanta, until we find all the energy of the universe packed in a few or even in a unique quantum.

Lemaître went so far to as to suggest what many specialists in general relativity now take for granted—that at some level space and time themselves *must be quantized:*

> Now, in atomic processes, the notions of space and time are no more than statistical notions: they fade out when applied to individual phenomena involving but a

small number of quanta. If the world has begun with a single quantum, the notions of space and time would altogether fail to have any meaning at the beginning; they would only begin to have a sensible meaning when the original quantum had been divided into a sufficient number of quanta. If this suggestion is correct, the beginning of the world happened a little before the beginning of space and time. I think that such a beginning of the world is far enough from the present order of Nature to be not repugnant at all.

Leaving aside the admitted philosophical confusion of talking about how something can be said to have come "before" time, Lemaître is clearly here grappling with the origin of all things—including space and time—from an initial quantum state. No scientist had ever done this before. He would later refer to this ultimate origin in his 1950 collection of essays *The Primeval Atom,* as "the now without a yesterday," which has been translated as "the day without yesterday," a quote often associated with Lemaître's letter to Nature.

Note that Lemaître was not talking about the "big bang" as it has come to be known—meaning, the hot explosion of all matter from pure energy. That concept would later be formulated in more detail by George Gamow, building on Lemaître's work. Lemaître was initially thinking of radioactive decay—from a super quantum sphere that divided and subdivided through a process of radioactive decay that gave birth to the evolving cosmos, a sort of cold big bang, as space expanded. For better or worse, Lemaître referred to this theory as the primeval atom. By 1931 it was already known that the atomic nucleus could not be treated as a single entity; atoms were not the smallest components of matter—they were made up of even smaller constituents, such as protons and electrons. Neutrons were discovered in 1932. Dirac's quantum field equations suggested the existence of the positron, the antimatter or counterparticle to the electron, a possibility that unnerved him, until it too was discovered by accident in 1932.

So even at this early stage, Lemaître might better have shied away from using the term "primeval atom." It would in later years cause no little amount of confusion and misunderstanding. "Primeval nucleus" would have been more accurate, and more difficult to confuse. "Primeval quantum" would have been even better, given Lemaître's letter to *Nature.* In any case, Lemaître proposed a super-dense state of already existing matter. It was cold, massive—indeed, containing all the mass of the known universe—and immediately began to disintegrate, its radioactive content literally forming the matter, time, and space out of which stars and galaxies and the cosmos would form.

There were some immediate problems besetting this thesis: just how could galaxies condense from the early expansion of matter? Could cosmic rays be determined to be the leftover fireworks, the "ashes and smoke" from the state of the cosmic origin? And forsaking Lemaître's earlier model meant there was another more serious problem his new primeval atom model faced, as Dirac related: the problem of the cosmic timescale suggested by the Hubble time, the age of the expanded universe, as it was then estimated (about two billion years).

Since Hubble had issued his results on the receding galaxies, it was determined that the universe was expanding at a certain rate, 525 km/sec per megaparsec. Assuming this rate was constant, it was possible to apply it to the most distant known nebulae and calculate backward, in effect, to determine the point from which the galaxies emerged, giving an estimate of when the universe began. This was known as the **Hubble time.** And based on Hubble's initial rate, it wasn't a lot of time. Stars were already estimated to be older than this, and for that matter, according to geologists, so was the planet Earth.

In his 1927 paper, Lemaître anticipated Hubble by estimating a rate of expansion not far from what Hubble would himself publish. And he further derived Hubble's law, that the velocity of recession for any galaxy is directly proportional to its distance, two years before Hubble did.

Although no documentation survives, it is very likely that Lemaître and Hubble were in correspondence on the subject of expansion in the years prior to Lemaître's 1927 paper as well as Hubble's own publication on the recession of the galaxies.[48] Lemaître was by his own admission a poor correspondent, meaning he was not organized about saving his letters. Hubble was just suspicious enough of anyone who might have claim to credit in anything he did, that he might not preserve any record of communication to undercut the significance of his

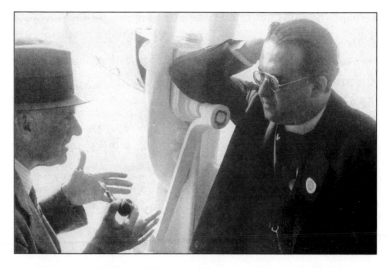

Lemaître on the boat to UK with Eddington, 1938—the last time they met in person. Photo courtesy of Archives Lemaître, Institut d'Astronomie et de Geophysique Georges Lemaître, Catholic University Louvain.

own contributions.[49] Nevertheless, many physicists have assumed there had to have been some contact between the two. Not that there was ever any contentiousness—for Lemaître certainly never complained about **Hubble's law** being named after his fellow astronomer.

In any case, the timescale of the expansion rate that Hubble estimated at the time was a problem for the primeval atom theory. It was not a problem for Lemaître's initial model of expansion, as it proceeded from an already preexisting state of static equilibrium going back to infinity. But once Lemaître theorized a temporal origin to his cosmic model, the Hubble expansion rate—assuming it was uniform—posed an immediate quandary. The expansion rate, extrapolated backward to a cosmic beginning, only afforded about two billion years as it was then understood. Obviously, the universe could not be younger than its constituent parts. But as problematic as this discrepancy was, it did not kill Lemaître's theory out of the gate.

Indeed within a short time, by 1948 in fact, other astronomers, notably Walter Baade, would reconsider Hubble's initial estimate and find it seriously flawed. Baade's recalculation, based on careful calibration of the period-luminosity relationship of two types of Cepheids in the Andromeda galaxy, doubled the estimated distance to Andromeda and also doubled the Hubble time, bringing the estimate of the expanded universe's age to about four billion years, which was more in line with stellar evolution and Earth's age. By the 1950s, Hubble's pupil and successor at Mount Palomar, Allan Sandage, would revise the age upward even more, so that the timescale would, in effect, no longer be a problematic issue for Lemaître's or any other big bang model.

In the meantime, Lemaître did not wait for astronomers to save his theory. He began revising his cosmic models in the mid 1930s, and it was here that Lemaître's daring (and fore-sight) became truly evident. He showed how the cosmological constant could be used as a force to speed up the expansion. His final model, which became known as the "hesitating" universe, originated from the super-dense state of the primeval atom, expanded quickly, then slowed almost to a halt for a requisite time before accelerating outward again.

In this way, Lemaître felt that Λ could help establish a model in tune with the Hubble time as it was then understood. The two-billion-year expansion period represented here only the latest expansion phase of the universe, rather than the entire evolution of it.

Although aspects of Lemaître's idea struck many as too ad hoc, his model actually matches better with what is known about the universe today than his initial model. Based on groundbreaking research from two teams of astronomers in 1998, which we will look at in more detail in chapter 9, it is now generally accepted that the universe is accelerating—the expansion rate has not been constant over the course of cosmic evolution, and the cosmological constant seems to be a decisive factor in its development.

Lemaître had been in America prior to his meeting with Einstein for some months, since August of 1932, when he traveled from Great Britain to Montreal. Circumstances from that visit afford the historian an engaging snapshot of the priest shortly after his arrival in Montreal, wandering the grounds of Hermitage Country Club in Magog, Quebec, not far from the U.S. border. Lemaître and a group from the Cambridge Observatory waited with others to see if they might catch a glimpse of the solar eclipse that took place on August 31. The Hermitage Club was one of the locations directly on the totality path of the eclipse. But Lemaître and his companions saw nothing due to the poor weather. Lemaître apparently found the situation ironic, since he had just heard about French astronomer Bernhard

Lyot's invention of the coronagraph, a telescopic instrument that allows astronomers to block out the disk of the sun in order to better view its corona, thus freeing them of their dependency on the vagaries of eclipse expeditions.

From Canada, Lemaître went to Cambridge, Massachusetts, once again. He and Eddington attended the Fourth General Assembly of the International Astronomical Union, and both were grilled by Shapley and others as to the details of this new "expanding model" of the universe and Lemaître's even stranger so-called fireworks theory of the cosmic origin. As with his first stay in Cambridge, Lemaître shuttled back and forth between the rivals of Harvard and MIT. Taking his inspiration from Shapley once again, he worked on a way to integrate observations of distant nebulae into his expanding model—as well as a means of reconciling the short timescale suggested by his initial model and Hubble's timescale. At MIT, Lemaître worked with his former advisor Manuel Vallarta on a theory of how cosmic rays might represent relics of the fireworks origin of the universe. By December he was on his way out to California again. In January, fortified with his semester's work back East, he gave the two seminars on his expanding universe model and cosmic rays to an audience at Cal Tech that included Einstein and was covered prominently by the press.

The publicity of his visit affords an interesting portrait of Lemaître at this time. The press, almost predictably, was more fascinated by the fact that Lemaître was a priest as well as a physicist, than they were by his actual theory:

And when the boy grew up and became both a priest and a scientist, he didn't believe that the whale swallowed Jonah or that the world was made in six days.

Science taught him that a whale could not survive the feat, and that creation took millions upon millions of years.[50]

The "boy" was now described as "a stoutish young man of thirty-eight who wears horn-rimmed glasses and the Roman collar of a secular Catholic priest. He begins with a smile and a French inflection in otherwise perfect English." And Duncan Aikman's article for the *New York Times* of February 19 is headlined, "Lemaître Follows Two Paths to Truth: The Famous Physicist, Who Is Also a Priest, Tells Why He Finds No Conflict Between Science and Religion."

As we already noted, there is some confusion as to the extent of Einstein's enthusiasm for Lemaître's primeval atom theory at this time. As Kragh points out, while the story in the March 11, 1933, issue of *Literary Digest* quotes Einstein as saying, "This is the most beautiful and satisfactory explanation of creation to which I have ever listened," it's more likely based on *Newsweek*'s January 23 report that Einstein was actually referring to Lemaître's "beautiful and satisfying interpretation of cosmic rays" as relics of the origin.[51]

Encouraging as Einstein was, it's unlikely that he regarded Lemaître's primeval atom theory as the last word on the subject—and unlikelier still that he would have employed the word "creation" to describe it. Einstein and de Sitter collaborated on a later model, for example, that avoided any temporal beginning altogether, which is one signal that—at least philosophically—Einstein preferred a model that did not require a temporal beginning to the world.

But what about the rest of the physics and astronomical community of scientists? How did they react to Lemaître's theory? They were certainly not nearly as exercised about it as the press was. It's obvious from the press coverage in 1933, that Lemaître was treated with the same sort of "celebrity" awe that followed Einstein. It's amusing, for example, to read him differently described as speaking in a thick Belgian accent in one account, to speaking in perfect English in another—"stoutish" with "horn-rimmed glasses," a scientific version of Chesterton's Father Brown.

It's noteworthy that when it was clear that Hubble's 1929 data supported the expanding relativistic models, Einstein decided to dump the cosmological constant for good. For Einstein, the constant was only a way to preserve the static nature of his model, which he believed had to be the only solution to his equations back in 1917. Once expansion was implied by Hubble's ever-increasing sample of galaxies, he wrote to his friend Paul Ehrenfest, saying, "Away with it!" Lemaître was not so sure. He believed the cosmological constant was more than just a mathematical balancing factor. He believed it represented an actual physical force—the vacuum energy of space, an energy that was essentially undetectable at the local level of, say, the solar system or nearby stars—but one that was very real on the scale of the galaxies and the universe as a whole. And to the end of his days he included Λ in his work—on the right side of the field equations, which govern energy density—underlying its physical connection to the energy density factors. Einstein placed the constant on the left side, signifying its purely geometric role, to prop up the static model. Once again, Lemaître's physical intuition prevailed over what Einstein felt was the mathematical inelegance of the lambda constant. And history has proven Lemaître more farsighted. The evidence since 1998, indicating that the expansion is accelerating, necessitates either a positive lambda, as Lemaître believed, or some type of similar energy force to explain it.[52]

But notwithstanding Einstein's dismissal of the cosmological constant, he was impressed with Lemaître's expanding model, once the observational data became widely known and Eddington publicized Lemaître's solution in 1931. Einstein in fact cited Lemaître's and Tolman's solutions when he visited Cal Tech two years before. Like de Sitter and many of his colleagues, Einstein embraced Lemaître's theory in light of the new data. He publicly accepted the new paradigm of expansion in April 1931 in Pasadena and at Mount Wilson, and mentioned Lemaître's work in conjunction with Tolman's as being the convincing factor for him.

Not long after they met in California in 1933, Einstein recommended Lemaître, alongside two others, for the Belgian Franqui Prize from King Leopold III, which—next to the Nobel Prize—was the most generous then in existence, giving Lemaître $33,000. The two would see each other again in the summer of 1933 under foreboding circumstances when it was clear Hitler was going to assume command of Germany. World War II was still six years away, but for Einstein the war with the Germans was already beginning. Einstein would never return home. Once Hitler was made chancellor, the Nazis seized Einstein's property and all of his papers on April 1. Later that month Einstein went to the German embassy in Brussels, where he relinquished his German passport. To make a clean break with his homeland in protest of Nazi policies toward the Jews, and perhaps to jog the conscience of German intellectuals who to this point were taking no stance against Hitler, he officially resigned from his professorships at the University of Berlin and the Prussian Academy of Science. He and his wife, Elsa, then settled for a time at Le Coq sur Mer on the Belgian coast.

Lemaître paid Einstein a visit shortly after he arrived, telling him that he had received support from the Franqui Foundation of Belgium to put together a series of seminars on spinors, essentially pairs of complex numbers that can be transformed through rotation—making them useful not only in general relativity, but also in quantum physics and multi-dimensional theories of space. Lemaître proposed that the seminars be given by Einstein. Einstein was very warm to the idea and agreed. Security was kept tight for the events, even though Einstein did not quite take seriously his wife's fear he could be stalked and assassinated in public. At her request, however, Belgian Queen Elizabeth did arrange for detectives to guard Einstein while he was in the country.

The seminars were held at the Fondation Universitaire in Brussels, beginning on May 3, where Einstein delivered the first of three presentations. After his last, Lemaître apparently couldn't resist suggesting different ways for Einstein to simplify some of the proofs he had presented. In response, puckish as ever, Einstein abruptly rose to tell the audience that the monsignor would have "some interesting things to tell us" at the next lecture. At first all but terrified by this challenge, Lemaître spent a weekend poring over his notes. But when the time came to make his presentation at the end of the conference on May 17, he was relieved and gratified at Einstein's many approving interruptions and comments of "*tres joli*" (very beautiful).

The Brussels seminars appear to be Lemaître and Einstein's last interaction together, although they did continue to correspond. It was not long before the founder of relativity was heading off to England and ultimately back to the United States, where he would spend the rest of his life pursuing his dream of a unified field theory at Princeton's Institute for Advanced Studies. Lemaître did spend a semester as a visiting lecturer there in the first half of 1935, where he likely met Einstein again, although there is no documentation of such meetings. When the shadow of war once again descended on Europe, Lemaître would find himself trapped and isolated in his own country (and at one point almost inadvertently killed by the Allies) once the Nazis invaded in 1940.

A long hiatus was on the horizon for Lemaître, though he did not suspect it in the midst of his travels in the 1930s, a long hiatus brought about by world events and personal events. These would conspire to stop him—almost completely—from further developing his cosmic theory. He had written to *Nature* and explained at the British Meeting in 1931 that progress on his "fireworks" theory of the universe would have to be made in quantum physics for the next stage of the cosmic origin's detailed development. That progress would be made just after World War II, spearheaded by one of the most unlikely scientific figures to arrive on the scene from behind the Iron Curtain—George Gamow.

7.

THE COSMIC MICROWAVE BACKGROUND

How seriously can one take a theory proposed by a reputed alcoholic and two scientists who have defected from academia to industry?

Helge Kragh, Cosmology and Controversy

Alpher notes that a number of his contacts with Gamow were at a bar and grill called Little Vienna, on Pennsylvania Avenue near George Washington University, usually for an hour or so before Gamow went off to lecture. The café is gone, but the memory of our meetings, and of the libations prior to the lectures, lingers on.

Ralph A. Alpher and Robert Herman, Genesis of the Big Bang

THE DEVELOPMENT OF cosmology took a curious turn after Lemaître's main contributions, during and after the lost decade of World War II. "Lost" here meant for many astrophysicists and cosmologists the time that was spent working on projects related to the war effort, or time spent in isolation from colleagues and from continuing work in their field. This was certainly the case for Lemaître, who not only was cut off from the rest of the free world by Germany's invasion of his homeland, but who also barely survived an American bombing raid that partially destroyed the apartment building where he lived.

Einstein had by this time left his native country for good and resettled in Princeton, New Jersey. There, in August 1939, informed by fellow physicists and exiles Leo Szilard and Eugene Wigner of the real possibility of producing weapons based on atomic fission, and alarmed by Hitler's belligerence (it was just a month before the invasion of Poland), he wrote his famous letter to President Franklin D. Roosevelt alerting him to such a possibility and the fear that Germany almost certainly would produce these weapons. Roosevelt would not reply until late October, and it would be over a year before secret funding for atomic bomb research would finally be granted.

Einstein made no more major contributions to questions of cosmology based on his theory—except to comment on a new paper or theory when it was brought to his attention (for example, by the irrepressible George Gamow, about whom more will be said). He continued to occupy himself with his unified field theory, unaware that the larger community of physicists was working its way shortly to an expanded knowledge of nuclear physics that would render Einstein's assumptions about gravitation and electromagnetism—as the two fundamental forces in nature—obsolete. He remained the center of a small coterie of students and colleagues at the Institute for Advanced Study in Princeton, where he worked in isolation.

Once the Germans began their march into France and—once again—into Belgium in 1940, Lemaître also underwent a trying period of retreat—not only because of the war and its effect on his homeland—but because of his family. He was not exactly caught by surprise by the German invasion. As early as May 1940 he and his parents tried to escape to the coast, the Pas de Calais, in order to cross the channel to England. But they were stopped by the German Panzer divisions assembling for Dunkirk, and forced to return to Brussels. Life under the Nazi regime was repressive, but not at first unduly harsh in terms of living conditions. The initial invasion was a shock; in fact, for the second time in his life the Germans burned Louvain's university library to the ground. But after briefly closing the university, the Germans soon allowed classes to resume.

Lemaître's father, Joseph, died in 1942, after a day of work like any other day for the seventy-five-year-old lawyer, when he collapsed on the trolley ride back home from the office. Joseph's death left Lemaître's mother alone in her house, and quite apart from the repressions imposed by the Third Reich,

Lemaître no longer felt the freedom to travel widely as he had done so much in the decade previously. As the oldest son, and not married, Lemaître felt it his responsibility to take care of her. This meant planting himself (with small expeditions excepted) for good in Belgium even after the Germans were driven out.

Two years later, in 1944, Arthur Stanley Eddington unexpectedly died from cancer after a quick decline. His last years, like Einstein's, were taken up with a grand mathematical project that struck many of his colleagues as more metaphysical than physical. Willem de Sitter, easily the eldest of the first generation of Einstein's cosmological protégés and whose work did so much to inspire Lemaître, had died suddenly of pneumonia in 1934 at the age of sixty-two. Cut off as he was, a continent away from the centers of astronomy and physics he had grown so used to visiting in his researches, Lemaître must have felt a sense that an entire world was slipping away forever.

Lemaître's withdrawal from the international scene and continued pursuit of the primeval atom theory was not all due to negative factors. His scientific interests continued to broaden. One was his growing fascination with mathematical computing. As early as 1933 he began using new, almost room-size computers to calculate solutions to problems in celestial mechanics for his classes. But he realized computers would also be helpful in cosmological problems as well, in particular the vexing problem of how galaxies, stars, and nebulae could coalesce out of the expanding matter of the universe during the later stages of expansion in the Lemaître model of the universe. As the years wore on he became more and more focused on computing and early "hacking" for its own sake, going so far as to help his university establish a center for electronic computing in the late 1950s. This no doubt distracted him from more rigorous development of his primeval atom theory.

Lemaître did reestablish contact with colleagues across the Atlantic after the war, but he did not accept any invitations to return to the United States, or indeed travel anywhere distant, until after his mother died in 1956. Lemaître declined an offer from Erwin Schrödinger to join the Institute for Advanced Studies at Princeton for a visiting professorship in 1951. This is doubly unfortunate in hindsight, considering that had he accepted the position he might have continued to prod Einstein in his few remaining years, about the advantages of the cosmological constant and perhaps discuss the cosmological applications of general relativity with Kurt Gödel, the famous mathematician whose late friendship with Einstein in Princeton led to his own fascinating contribution to cosmology.[53] Lemaître also turned down an offer from the European Division of the U.S. Air Force Office for Scientific Research when they asked him to collaborate in recruiting a team to study the data from satellites that had detected the Van Allen belt around the Earth.

Lemaître's long-time colleague and later biographer, Odon Godart. Godart tried in vain to convince Lemaître to collaborate with Gamow on his revised "hot" big bang theory of the late 1940s. Photo courtesy of Archives Lemaître, Institut d'Astronomie et de Geophysique Georges Lemaître, Catholic University Louvain.

Although he wrote no new papers elaborating or modifying his primeval atom theory after 1934, after the war Lemaître did collect a number of general essays and lectures he had written on the subject, and they were published in 1950 in French under the title *L'Hypothese de l'Atome Primitif.* The English translation, *The Primeval Atom: An Essay on Cosmogony,* appeared the same year in the United States and Britain. (The book has long been out of print.)

Lemaître's theory of cosmic rays, developed with Manuel Vallarta, as the leftovers or fading fireworks of the original cosmic explosion, did not attract any support. Indeed, as early as 1938, the notion that cosmic rays could be the remnants of a primeval nucleus was undermined by the research of Arthur H. Compton and Carl D. Anderson, reported at a 1938 symposium at Notre Dame University that Lemaître and Vallarta attended. Somehow both missed the importance of Compton and Anderson's work.

Lemaître's initial model of expansion, his primeval atom, suggested a cosmos of an age roughly two billion years, as noted earlier. This incorporated the Hubble time, as then estimated by Hubble and Humason based on their measurement of redshifts and estimates of distances between galaxies (as accurately as possible at that time). This expansion scenario conflicted with what was estimated for the age of the earth from geological and radioactive evidence, and for the age of the sun and stars based on the increasingly detailed models of stellar evolution being worked out by nuclear physicists and astrophysicists such as George Gamow and Hans Bethe. And Lemaître's insistence on the cosmological constant as a physical factor in helping explain the apparently short Hubble time's conflict with the more accepted view of the Earth's and stars' ages also worked against his theory, as it was viewed as too ad hoc. Lemaître's primeval atom theory thus seemed to be in a state of abeyance and in danger of being completely dismissed. The real issue, Lemaître knew, as he said back in 1931 at the Royal Society Conference, was accounting for the early stages of the cosmic expansion in more detail and how it could lead to the development of stars and galaxies. Lemaître knew the logical next step of his theory's development had to be based in using quantum physics to describe the early nuclear stage of his fireworks universe.

Also of keen interest to him was the need to explain how the star clusters and galaxies coalesced out of the initial stages of cosmic expansion, an issue that the ever-nagging Harlow Shapley never ceased to raise whenever he and Lemaître met. Lemaître believed that continued exploration of the field equations with a positive cosmological constant would suggest an answer. But he did not himself pursue any theoretical investigation of how nuclear physics would help expand his theory. Throughout the 1930s and after the war, he seemed content to leave this side of the problem to others. In fact, by the mid 1930s the application of nuclear physics to the evolution of the universe was already under way. But not in Europe. Oddly, during the years leading up to and during the war, American physicists for the first time took the lead in continued development of the expanding big bang universe, and their contributions would significantly alter Lemaître's model in ways he had not anticipated. This was due largely to the work of George Gamow.

Beginning in 1938, the Ukrainian-born nuclear physicist with a special connection to the expanding universe models took over the field. Gamow was born in Odessa and was educated in Russia just as the czarist regime gave way to the new Leninist state. While Gamow was still studying in Moscow in 1923–1924, he attended the lectures of Alexander Friedmann and heard firsthand about the possibility of time-dependent models of Einstein's universe from the mathematician who first developed them. This was at about the time Lemaître himself was studying general relativity in Belgium. In his autobiography, *My World Line,* Gamow said that he intended from the outset of his studies to specialize in theoretical physics and general relativity, but that Friedmann's untimely death in 1925 forced him to transfer from Friedmann's encouraging tutelage to that of another, less inspiring physicist, one who was not interested in general relativity.[54] Gamow by fate had to concentrate more on nuclear physics. It was a fortunate fate, though he didn't know it at the time. Gamow's background in nuclear physics would give him the tools he needed to delve where Lemaître wouldn't.

Gamow studied at Göttingen for the summer of 1928, before Niels Bohr invited him to the Institute of Theoretical Physics at the University of Copenhagen for a yearlong fellowship. At the time, Bohr was preoccupied with mapping out his interpretation of the new physics that would cause Einstein so much distress. But he was impressed by Gamow, who used Schrödinger's wave equations for quantum physics to give a quantitative description of alpha radioactivity . . . how the alpha particle—which is essentially a helium atom, two protons and two neutrons, stripped of its electrons—could "tunnel" its way out of the nucleus (where it was believed to reside) although it had less energy than was necessary in classical terms to escape. This theory brought Gamow some attention, and he soon learned that an inverse application of this "tunneling" effect might be useful in trying to figure out how the atomic elements are built by nuclear reactions inside stars.

For Gamow it all came back to the stars. In the 1930s he and many other physicists were interested in something Arthur Eddington had done much to describe: how stars generate their energy and how the elements are formed. Eddington had first suggested that stars generate their energy via Einstein's famous $E = mc^2$ equation. This seemed to offer an explanation as to how the stars could shine for so long, and it stood in stark contrast to Lord Kelvin's nineteenth-century theory, which argued that stars shone based entirely on energy created by their gravitational contraction alone. The problem with Kelvin's theory was that a contraction-based system did not allow for stellar lifetimes longer than a few hundred million years, which plainly conflicted with known evidence of the earth's age, as well as the sun's.

In 1933, now in the United States after he and his wife fled the Soviet Union, Gamow began writing papers on stellar physics. He was particularly interested in trying to figure out how to use nuclear processes to explain the relative abundances of the heavier elements. Within a few years he was arguing that the recently discovered neutron played a significant role. "The neutrons which can be ejected from the nuclei of light elements by collisions with protons," he wrote in 1935, "may stick to the nuclei of different heavy elements thus securing the possibility of the formation of still heavier nuclei." Gamow's next step was trying to describe how such reactions might take place in the interior of stars. This wasn't easy. Beyond the dramatic and secret work on the Manhattan Project (which Gamow at first was excluded from because of his brief service in the Red Army before fleeing the USSR), many American scientists were drafted to support the development of radar, weapons, and defense for the U.S. Navy, Air Force, and Army. The later benefits of technological development during the war for astronomy and astrophysics, for the testing of Einstein's general theory of relativity, were significant and would soon make themselves apparent after the war was over. But for the duration of the war, nuclear physics, not astrophysics, predominated.

As an outsider, not immediately welcome to the war effort, Gamow was free to take the lead in organizing conferences and meetings to interest others in the role of nuclear physics in the evolution of the early universe. He started to do so in 1938, when he organized a conference in Washington, D.C., on the subject of nuclear physics and the development of heavier elements (meaning anything more complex than hydrogen and helium, the universe's most common elements). The 1938 conference was called "Problem of Stellar Energy-Sources," fourth in a series of conferences arranged jointly by George Washington University and the Department of Terrestrial Magnetism of the Carnegie Institution. Thirty-four scientists attended—astrophysicists and nuclear and quantum physicists. Hans Bethe and Edward Teller, who had prominent roles in the Manhattan Project, were among them.

Lemaître was in America at the time, too—at Notre Dame, as a visiting lecturer, as noted above. Just a few weeks after Gamow's conference, he attended the Notre Dame conference on "Physics of the Universe and the Nature of

Primordial Particles," which more than one hundred scientists attended. Lemaître delivered his talk on density fluctuations measured by both Shapley and Hubble, and how he could develop relativistic cosmological models to account for them. As mentioned before, he missed the significance of the report discounting cosmic rays as primordial artifacts of the primeval atom. In hindsight it is ironic too that Lemaître did not hear of Gamow's Washington conference, or that if he did, he did not see the subject matter as worth investigating, given its relation to his interest in the formation of galaxies and its importance for his theory.

One scientist who did not miss the significance of Gamow's conference on nuclear physics of stars was soon to make a groundbreaking discovery. In 1939, a year after the conference, Hans Bethe published his paper on the so-called CNO cycle—the carbon-nitrogen-oxygen cycle—theoretically modeling in detail how stars turn hydrogen into helium by fusing elements such as carbon and nitrogen and generating a great deal of energy, a paper for which he later won the Nobel Prize.

But neither Bethe nor anyone else could take this process further to account for the formation of heavier elements inside stars. Indeed, he could not account for how the sun had carbon in the first place. As a result, Gamow believed that a cosmological cause, prior to the formation of stars and galaxies, was the answer, and that it was connected to the expansion of the universe. It would be a few years, however, before Gamow could make further progress. Although his prewar 1938 conference went well, the next one Gamow convened, in 1942, was less of a success, mainly because of the war. With more pressing problems facing the majority of physicists, how the universe evolved seemed of small interest.

Science does not progress by consistent logical steps and insights. Often the "wrong" theory will inspire a scientist to the "right" answer. Indeed, Gamow worked his way toward his big bang model by making a vital but wrong assumption at the time: that the heavier elements in the universe, such as iron, lithium, and uranium, were forged in the beginning stages of the cosmos, long before stars themselves were formed. This was an inspired mistake. As we've seen, papers such as Bethe's convinced astrophysicists that stars were incapable— during the normal course of their existence—of producing the heavier chemical elements. They had to be produced in some other fashion. Gamow decided to figure out how the early universe might have been responsible. This was before the realization of the importance of supernovae, and how their cataclysmic explosions fused the heavier elements. Taking his cue from Bethe's paper, Gamow started working on the problem of how the simplest elements, hydrogen and helium, could have developed in the early stages of the Lemaître cosmos, before any stars existed. It was his hope that a detailed theory could provide the next step to showing how the rest of the elements were forged as well.

The key here was his belief that Lemaître's primeval atom was not an atom but more like a huge, cold primeval nucleus—one consisting entirely of neutrons (which were a "hot" subject at this time, having just been discovered). Neutrons were known to have short lives outside the nucleus of an atom and would decay into protons and electrons. Gamow believed there had to have been a stage in the early universe when the great nucleus began to decay, just as Lemaître believed, but in such a way that the freed protons and electrons in the super-dense early stage of the universe could be fused under pressure into many of the heavy elements that formed the later universe. This struck many of Gamow's colleagues as fantastic. Gamow was a flamboyant character, given to pranks and sometimes truly outlandish ideas. According to British astronomer Fred Hoyle, whose steady state theory was a rival to Gamow's: "I do not recall reading any paper by Gamow that wasn't short. When you knew Gamow you knew why his papers were short. This was because George couldn't concentrate on anything for very long. After he had talked about some issue or other for ten minutes he would pull out a pack of playing cards or a box of matches from a coat pocket, proceeding to perform some trick or other which interested him more than it interested you."[55]

Although he was known as a sloppy mathematician, Gamow had a fertile imagination, and his enthusiasm attracted talented graduates. Among them were Ralph Alpher and eventually Robert Herman, both sons of Russian Jewish immigrants—and both talented graduates of New York's City College who wanted to make a name for themselves. Gamow set Alpher the task of "working out the development of structure in the universe, requiring calculations of the behavior of various kinds of density perturbations in a relativistic, homogeneous, isotropic expanding universe containing matter only." This Alpher did, laboriously, coming to the conclusion that "small perturbations of the density could grow but not on any cosmologically useful time scale." In Alpher and Herman's account many years later, they recalled:

> In late 1946, unfortunately for us, Gamow received a then current issue of the Journal of Theoretical Physics of the USSR containing a paper (also a Ph.D. dissertation) by Evgeny Lifschitz on the same subject and with the same conclusions. Alpher has a vivid recollection of Gamow coming into his office, waving a copy of the journal, and saying, "Ralph, you have been scooped." In one of a number of silly things Alpher has done, he destroyed his voluminous notes on this subject (with pages and pages of perturbed Riemann-Christoffel brackets) and decided, with Gamow, on a second dissertation topic: developing Gamow's rather cursory 1946 ideas on primordial nucleosynthesis in the early stages of an expanding universe.[56]

Specifically, this meant working out how hydrogen and helium could be developed out of the decaying neutron gas of the early phase of the universe. This led to Alpher and Gamow's 1948 paper, also known as the $\alpha\beta\gamma$ paper, a paper that has become a milestone in big bang history—and one that, like Lemaître's and Friedmann's important papers, was almost entirely overlooked at the time.[57]

It's important to note that there is no direct succession between the work of Lemaître and that of Gamow (although it appears there could have been, and Gamow was certainly aware of Lemaître's work). Nuclear physics was Gamow's inspiration here—not at first the influence of the prior mathematical work of Lemaître or even Gamow's first teacher, Friedmann. Gamow had read Lemaître's, de Sitter's, and Eddington's papers on expanding relativistic models, but initially he did not apply them to his physics of the early universe. It was only later in the 1940s—1948 to be precise, with the $\alpha\beta\gamma$ paper—that Gamow and Alpher and Herman realized that the model they wanted needed to originate in the Lemaître model—and more importantly, it had to originate in a hot state, not in a cold nucleus as Lemaître had envisaged. Only a hot state of millions of degrees, they reasoned, could allow nucleosynthesis to "cook" elements like hydrogen, helium, and heavier elements. Gamow, in other words, took Lemaître's primeval atom and turned it into the recognizable big bang model that remains the base of the standard model to this day. One immediate consequence of such a hot big bang model, Gamow and his team realized, is that radiation from the primeval fireball *should still remain,* albeit at very attenuated wavelengths in the radio end of the electromagnetic spectrum. Where Lemaître's atom model led him to examine cosmic rays as candidates for the leftover fireworks, Gamow's model more simply suggested a low hum of microwaves in the background of the universe. Alpher calculated a temperature to this background radiation of about five degrees Kelvin.

At this point we are confronted with another puzzle. Although Gamow and Alpher did point out that the five-degree radiation should be measurable, they did not specifically suggest an experiment to detect it—although the equipment existed at the time to do so. In fact, the background radiation had *already been detected* for the first time almost a decade previously.[58] But in their follow-up papers Alpher and Herman did not ever mention or seem to be aware of them. Still, they did not specifically challenge the wider community to test for this radiation, although the radio technology was now available. This was an oversight. Given the topics in nuclear physics and astronomy that were then dominating, Gamow, Alpher, and Herman did not gain much attention. History would not be kind to them. It would be a full fifteen years before an entirely different team essentially rewrote the same paper, but with an emphasis on probing for the background radiation, and this team in its own turn would be beaten to the actual discovery by two Bell Lab physicists—by accident.

In 1965 Robert H. Dicke and P. J. E. Peebles and their team at Princeton worked out the details of a background radiation based on expansion from a fireball state of the early universe. They knew nothing of the work of Gamow, Alpher, and Herman. Indeed, Dicke was motivated primarily by an interest in oscillating models of the universe, and his suggestion that the renewed expansion phase after a cosmic collapse would result in a universal black-body radiation. He and Peebles predicted such a background noise would measure between three and five degrees above absolute zero. They built a radio detector to measure it themselves, but found that someone else had come across the background noise first. Two physicists working for Bell Labs, Arno Penzias and Robert Wilson, had been preparing a radiometer in New Jersey. Dicke and Peebles were elated; Penzias and Wilson were less so, as they had always assumed the "noise" was an unwanted interference from their real goal of studying radio sources among the stars.

Once the two teams communicated, they agreed to write separate papers about the discovery. Dicke and his team published their letter on the prediction of a **cosmic microwave background radiation** in the same issue that its accidental discovers, Penzias and Wilson, published theirs. This was the groundbreaking discovery of what more flamboyant journalists and writers have called the "echo of creation." This echo, by the way, can be "seen" by anyone with a television and antenna, unplugged from any cable source, by turning to a blank channel. Some of the snow on the static screen is caused by the cosmic microwave background radiation—about as close as science can get to revealing the existence of a ghost in everyday life.

Penzias and Wilson were awarded the Nobel Prize for the discovery in 1978. Dicke and Peebles were not recognized, even though they rigorously recovered the theoretical ground that Gamow, Alpher, and Herman had worked out in such detail a decade before. Gamow was reportedly bitter about this, and after an apology and some excuses, Dicke duly gave credit to the originators once they reviewed the 1948 paper. If Dicke and Peebles were bitter about being overlooked by the Nobel committee, Alpher and Herman were even more so:

> We do not accept the argument of some that correct attribution does not matter, but that only the further-ance of science matters. This view does not reflect the ideals and realities of the scientific enterprise. A correct history of science as a human endeavor does matter. . . . For many years we have contemplated the possibility that the reason for a lack of acceptance of the worth of our early work lay in our both being employed by large industrial research laboratories at the time when the background radiation was first observed, and for some years thereafter.[59]

As for Lemaître, he was delighted when he finally heard the news (almost a year after the discovery). Many years had passed since he first suggested the primeval quantum in his letter to *Nature* in 1931. When the hardest evidence for the big bang theory was discovered, the founder of modern cosmology, Albert Einstein, had been dead for a decade. Georges Lemaître was only a week or two away from his own death, having suffered for several weeks from leukemia. His old assistant and colleague Odon Godart brought the news to his bedside in June of 1966. Gamow himself was just two years away from an untimely demise from cirrhosis of the liver.

Although he wrote a popular book on the subject of the universe in 1952, professionally Gamow soon drifted away from cosmology just as Lemaître and Einstein had before him. Yet his contribution to the big bang theory is key. In many ways, he gave Lemaître's primeval atom a makeover, from the standpoint of nuclear physics, and he and his protégés Alpher and Herman provided the crucial prediction of the leftover radiation from the cosmic explosion that would lead to the theory's acceptance by the majority of astronomers and physicists in the 1960s when the cosmic microwave background radiation was discovered.

What's fascinating is that as Gamow's papers gained attention in the 1940s, Lemaître's colleague and later biographer, Odon Godart, suggested more than once that Lemaître contact Gamow and collaborate on a more detailed theory. Godart remembered that Lemaître always refused.[60] Why? Perhaps by this time, Lemaître had either lost interest, or felt that there really wasn't any more to be done with his theory.[61] He was also fully engaged in the use of computers for mathematical computation of problems in celestial mechanics. It may be possible that the reason Lemaître did not bother is that for him the whole question had been and always was a question of general relativity and the model itself—not the physics. With the advent of Gamow and his team, big bang physics became mainly the domain of American physicists and astrophysicists. The Europeans had been more interested in the mathematical geometric structure of the cosmos according to general relativity than the more pragmatic physical aspects of its initial constituents. There may also have been a personal component for his reason not to contact Gamow, which will be examined in more detail in chapter 11.

Whatever the case, Lemaître did not take Godart's advice, and the chance for a collaboration between Gamow and Lemaître in the late 1940s and early 1950s slipped away. This is unfortunate, given what happened to Alpher and Herman, who by the mid 1950s had turned to careers in industry. Gamow, his interest moving to the new, exciting field of DNA research, also drifted away from actively promoting the big bang theory and the prediction of a background radiation. More tragically, he was succumbing during this period of years, to alcoholism, a problem that led to growing isolation from students and colleagues and eventually killed him at the age of sixty-four. It's possible that a collaboration or at least a regular correspondence with Lemaître during the 1940s and 1950s might have reinvigorated both of their commitments to the big bang theory.[62]

In the end, it didn't happen. Interest in Gamow, Alpher, and Herman's famous paper of 1948 was—at the time—low. Their prediction of a five-degree Kelvin background radiation was ignored. That same year, Richard Feynman gained attention for his theory of quantum electrodynamics, and Edwin Hubble began his last much-publicized stint on the new two-hundred-inch telescope at Mount Palomar.

Gamow's bold prediction of a background radiation went almost unnoticed.

Meanwhile a competing theory of cosmology, formulated by Fred Hoyle and his collaborators in Britain, gained the support of physicists who were uncomfortable with the primeval atom and indeed any model of the universe whose evolution implied a temporal beginning of the world. The steady state theory, which is examined next, faded with the discovery of the background radiation—but not before spearheading a program of research and discovery of its own that would galvanize the astrophysics of stellar evolution even as new discoveries turned Einstein's general theory of relativity from an obscure branch of specialized mathematics into one of the hottest subjects in physics.

THE INSTITUTE FOR ADVANCED STUDY
Founded by Mr. Louis Bamberger and Mrs. Felix Fuld
PRINCETON, NEW JERSEY

September 26, 1947

Professor G. Lemaitre
9 rue Henry de Braekeleer
Brussels, Belgium

Dear Professor Lemaitre:

I thank you very much for your kind letter of July 30th. In the meantime I received from Professor Schillpp your interesting paper for his book. I doubt that anybody has so carefully studied the cosmological implications of the theory of relativity as you have. I can also understand that in the shortness of T_0 there exists a reason to try bold extrapolations and hypotheses to avoid contradiction with facts. It is true that the introduction of the λ term offers a possibility, it may even be that it is the right one.

Since I have introduced this term I had always a bad conscience. But at that time I could see no other possibility to deal with the fact of the existence of a finite mean density of matter. I found it very ugly indeed that the field law of gravitation should be composed of two logically independent terms which are connected by addition. About the justification of such feelings concerning logical simplicity it is difficult to argue. I cannot help to feel it strongly and I am unable to believe that such an ugly thing should be realized in nature.

First page of Einstein's letter to Lemaître in September 1947, regarding the cosmological constant. Image courtesy of Albert Einstein Achives, Jewish National and University Library.

8.

LEAN YEARS

EINSTEIN'S GENERAL THEORY of relativity has been such a fertile field of research for physicists for so long—the last four decades—that it's easy to overlook how little importance it held for the first four decades of its existence, prior to 1960. The three initial tests of the theory—the bending of starlight, the advance of the perihelion of Mercury, and the redshifting of starlight—were enough to satisfy the specialists who knew and studied the theory in its first years.[63] This was decidedly not the case for physicists who were interested in practical applications. Outside of that small coterie of cosmologists (including Lemaître and Eddington) there was little attention paid to general relativity. Physicists were interested in atomic physics, elementary particles, and those getting caught up in the science of rockets and space exploration found Newtonian mechanics perfectly sufficient for their needs. Indeed, up until the late 1950s, general relativity wasn't even taught in physics departments as a rule; it was taught in mathematics departments, where it belonged, as far as most physicists were concerned.

Einstein himself did not invest any further energy into the theory. As far as he was concerned, general relativity was but a step toward his more ambitious and quixotic attempt to unify the electromagnetic and gravitational fields into one cohesive theory. In 1942, for example, he wrote in the introduction to Peter G. Bergmann's *Introduction to the Theory of Relativity* that general relativity "has played a rather modest role in the correlation of empirical facts so far."[64] But he took no interest in new tests. This is not to say that there weren't any. Before the 1960s, Jean Eisenstaedt writes in his review of the period, several attempts were made to apply general relativity to problems such as the perihelion of the earth, the secular acceleration of the moon, the displacement of the orbit of Mars, and the influence of gravitation on atomic energy levels. But the expected results were so subtle that the experiments were not considered convincing. J. Robert Oppenheimer went so far as to note that in the forty years that had elapsed since the theory's inception, the three main tests "have remained the principal and, with one exception, the only connection between the general theory and experience." This comes from the man who first modeled the concept of black holes in 1939, as we shall see, using Lemaître's modification of the Schwarzschild solution to the field equations.[65]

All this would change by the mid to late 1950s, when technology began to suggest new, more rigorous ways to test the theory—when lasers could be used, for example, to measure its subtle effects. Radar, which had been developed for military purposes during the war, gave birth to radio telescopes, which took on a more prominent role in astronomy (and soon led to the discovery of quasars and pulsars), raising new questions not only about the evolution of the universe but also about some of the increasingly exotic objects in its farthest reaches.

For the period during and immediately after the war, cosmology remained general relativity's most fruitful, if speculative, application. Gamow, Alpher, and Herman were virtually the only physicists in the United States using the general relativity field equations to model their physics of the early universe. But in the United Kingdom, Fred Hoyle began to wonder how the relativistic equations could fit into an unchanging, steady state model of the universe. He would ultimately partner with Hermann Bondi and Thomas Gold to work out in detail a counter theory to the big bang.

Prior to both of these teams, however, tantalizing evidence of the big bang, specifically what would be known as the cosmic microwave background radiation, had already been discovered, though not by design. In Australia, two astrophysicists, Walter Adams and Andrew McKellar, first took the temperature, in a manner of speaking, of outer space, as early as 1937 and 1941, respectively. Their figure of 2.3 degrees Kelvin was the temperature of cyanogen molecules detected with a microwave receiver in a restricted part of the universe. They were not looking for evidence of a universal background radiation and believed the temperature they detected was due to the excitation of the molecules from interstellar radiation. They did realize they were measuring a sort of thermal bath of radiation, but they did not attribute it to any cosmic radiation left over from Lemaître's fireworks universe. Neither did anyone else.

Although their results were known to many physicists, Gamow and his colleagues were never aware of these findings. Hoyle would discover them, but would not really grasp their significance given his stake in an alternate, nonevolving theory of the universe. Only in hindsight, after the 1965 discovery by Penzias and Wilson, would the significance of Adams and McKellar's findings be realized.

Many accounts of the big bang's history tend to focus on the rivalry between the theory and its chief opponent, the steady state theory, often concluding with the steady state's "demise" due to the discovery of the cosmic microwave background radiation. However, the truth is that the steady state theory was never really considered much of a rival to the big bang to physicists outside the United Kingdom, where it was conceived. Nor was Gamow's big bang considered the established view, although some form of the expanding model, for instance the **Eddington-Lemaître model,** was favored by most astronomers and physicists. Despite its being ultimately discredited by the discovery of the cosmic background radiation, the steady state theory nevertheless inspired a great deal of fruitful research that extended and expanded the development of the big bang theory itself.

None of this would've happened had it not been for the late Fred Hoyle, the steady state's chief proponent and the big bang's most persistent critic. (It was Hoyle who first coined the term "big bang" to disparage the theory, in a 1950 series of lectures for British radio.[66]) Born in Yorkshire in 1915, Hoyle was as irascible as George Gamow was irrepressible and enthusiastic. And although the two disagreed about the state of the universe, they got along well. In fact, at Gamow's prodding, Hoyle contributed to some key papers on the big bang theory in spite of his misgivings about it. This speaks volumes for his open-mindedness as a scientist, considering how curmudgeonly he is often portrayed. (It's also interesting to note that in spite of his disdain for Catholicism and a certain streak of anticlericalism, he also got along well with Lemaître, as exemplified by a little-known span of two weeks that Hoyle and his wife spent together with the cleric when Lemaître accompanied them for a driving tour through northern Italy in 1957.[67])

Just as Gamow inspired Alpher and Herman to detail the big bang model as it came to be known, Hoyle, who was employed by the United Kingdom to develop radar during World War II, met and encouraged Gold and Bondi, both Jewish transplants from Austria, to work out their counter theory. In 1947–1948, the same period that Gamow and

Alpher worked out their $\alpha\beta\gamma$ paper, Hoyle, Gold, and Bondi developed, in separate papers, their theory of an unchanging universe—one that also expanded according to the equations of general relativity, but one that did not have a temporal beginning in the remote past—a universe, in a sense like Aristotle's unchanging firmament, that did not evolve.

There were some differences between them at first. Hoyle was familiar with the Friedmann-Lemaître equations, and he wished to build his theory around the equations even as he resisted the implication that an expanding universe had to have evolved from an origin. Hoyle in fact based his interpretation on the de Sitter model, an empty model that expanded, but the difference being that Hoyle's model included mass and energy. Gold and Bondi, on the other hand, objected strongly to the use of general relativity equations, questioning their perceived applicability to the universe as a whole. They built their model in much less mathematical terms, around the continuous creation of matter in space to sustain an eternal cosmos.

The key to the steady state theory was the rejection of the expanding universe from a prior state of super-dense matter. While Hoyle and company accepted Hubble's findings of galaxies in recession and agreed with the interpretation of expansion overall, Hoyle, Bondi, and Gold believed that a fine, continuous production of hydrogen atoms throughout the universe was enough to feed the expansion of space and sustain an eternal universe forever. In fact, according to Hoyle's calculations, the spontaneous creation of one hydrogen atom per ten billion cubic meters of space per year was enough to account for the then-currently estimated mean density of the universe, a factor they believed was strongly in the theory's favor.

But a fine continuous creation of matter out in the middle of empty space on its face meant violating one of physics' precious conservation laws: the conservation of mass/energy. Hoyle, Bondi, and Gold did not think this was too much of a stretch, since in principle such a process could be explained using quantum mechanics and therefore might one day be observable. The same, they argued, certainly could not be said of the initial explosion of the universe from a singularity. They had a point. The steady state theory also had the advantage of avoiding the depressing "heat death" scenario that Eddington had described earlier. The universe would not slowly wind down and die with a whimper, the way expanding universe models based on the Friedmann-Lemaître equations suggested. Hoyle further argued, based on his other work, that the heavier atomic elements, so key to the evolution of galaxies, stars, and ultimately life, were forged inside stars, so there was no need to rely on an early intense phase of cosmic evolution to explain the creation of heavy elements— as Gamow then believed.

But for all these solutions, there was a gap between the creation of hydrogen atoms in empty space and the fusion of heavy elements inside the bowels of stars. What the steady state theory didn't do was account for the cosmic abundance of the second lightest element, helium. Hoyle admitted this was a problem, but not a fatal one. The theory wasn't received well at first. Or rather, it didn't draw much attention at all, until Hoyle decided to talk it up outside of strictly scientific circles during his radio lectures on cosmology in 1950. This annoyed a great many scientists but did get the theory the attention it had thus far failed to attract. Most scientists, including Lemaître, took exception to its violating the laws of the conservation of energy in order to produce new hydrogen atoms out of nothing. American scientists were also put off by what they perceived as the British tendency to place too much emphasis on the importance of a priori principles in theory. As American astronomer Allan Sandage later recalled, Bondi went so far as to suggest, in terms considered somewhat heretical for a scientist at that time, "that whenever there is a conflict between a well-established theory and observations, it's always the observations that are wrong. And when he announced the first theorem in England and it was read by the Mount Wilson astronomers, they just dismissed all the steady-state boys. That was the beginning of the rejection. Bondi was associated with Gold and Hoyle, and they made such outrageous statements that they just couldn't be believed."[68]

It's been tempting for historians to assume Hoyle, Bondi, and Gold were motivated to develop the steady state theory purely by their collective atheism and dislike of the notion of a special "event" implying creation in the past. Hoyle once rather disingenuously said, "It is not a point in support of this [steady state] theory that it contains conclusions for which we might happen to have an emotional preference." And in his popular writings—for example, his 1950 *Nature of the Universe*—he did not hesitate to associate his atheism with the steady state model. And while this probably cost him the support of many scientists who otherwise might have been more supportive of the theory, neither he, nor Gold, nor Bondi ever let their opinions influence their papers. They did not conceive the idea of hydrogen atoms being created out of nothing in order to sustain the steady state, as has been suggested in many books. Rather, they concluded based on their model and its exigencies that some kind of matter creation had to be responsible, and they were not the first physicists by any means to suggest the continuous creation of matter in the universe at the time.

Likewise, the motivation from religious/philosophical conviction has also been true of treatments of Lemaître, who has been assumed in his work to have been ultimately inspired by his Catholicism to look for a "Genesis" moment in his fireworks theory.[69] As we saw in chapter 6, Lemaître had a perfectly reasonable scientific motive for "winding" his relativistic model backward to a starting point in time: his conclusion that the static Einstein model, the initial base from which his first expanding universe model evolved, could not last in a temporally infinite state—its own perturbations would severely undercut its stability. So the universe somehow had to have evolved from an initial state, or singularity. Penrose and Hawking's work in the 1960s proved him to be correct—any model of the expanding universe based on the relativistic field equations must have its source in a singularity.

Hoyle had his own reasons for disliking the Friedmann-Lemaître models: he believed that an evolving universe implied that the laws of physics themselves might have evolved, that they might have changed over time. For Hoyle this implied that the very laws of nature could not be seen or depended on as truly universal. He objected to this on a priori grounds and believed that the only universe in which the laws of nature could be seen as unchanging and stable had itself to be unchanging—not time-dependent.

Alone among the initial audience to hear about Hoyle's theory, Irish-born William H. McCrea, a professor of mathematics at Royal Holloway College at the University of London, supported it, because he thought the steady state theory was a better candidate than the evolving models to explain the large-scale structure of the universe. McCrea believed that the rate of continuous "creation" of matter, which could probably be explained at the quantum level in the near future, determined this. While Hoyle, Bondi, and Gold welcomed his support, they did fret somewhat when McCrea ventured his opinion that his preference for the theory satisfied his own, overtly Christian philosophy.[70]

Gamow seems to have dismissed the theory out of hand without really bothering to study it in any detail. Hoyle, who carefully read and reviewed Gamow's book, *Theory of Atomic Nucleus and Nuclear Energy Sources,* cowritten with C. L. Critchfield in 1950, was apparently put off by Gamow's miscalculation of the cosmic background radiation at ninety degrees Kelvin and did not look into the idea more closely.

This is especially ironic, since, had Hoyle done so, he might have taken more seriously the connection to the work of Adams and McKellar, whom he wrote had a more likely figure. Again, a chance for the importance of the cosmic microwave background radiation to be recognized passed.

While Gamow may have been aware of Hoyle's generally positive review of his book, Alpher and Herman apparently were not. Otherwise they might have taken notice of the findings of Adams and McKellar, which Hoyle alluded to, and they might have drawn more attention to their own cosmic background prediction. This in turn might have spurred some experimental search for it fifteen years before it was eventually discovered. As for Einstein, his only reference to the steady state theory (in 1952) was dismissive. In his 1962 book, *Einstein: Profile of the Man,* Peter Michelmore said Einstein described it to a young acquaintance as nothing more than a "romantic speculation."[71]

Lemaître did not agree with the steady state, either. For one thing, although he understood the reason for the theory's violation of the conservation of energy in order to pop hydrogen atoms out of nothingness to stabilize the eternal universe, and that it might be explained by quantum mechanics, he did not see why this one modification of the conservation laws should be singled out—and not others as well. If conservation of energy should be altered for the sake of fitting a theory, for example, why stop there? Why not modify the other principles if they are not convenient to the theory? But if he had such thoughts, Lemaître never wrote about them. And whatever disagreements he had with Hoyle, Gold, and Bondi on the subject were entirely technical, never personal.

Indeed, for all the legends about the contest between the two theories, there is no evidence the players on either side had anything but personal respect for each other (although Gamow did playfully mock Hoyle in a satiric poem about the steady state theory in his 1950 book, *Creation of the Universe*). Hermann Bondi, for example, recalled a pleasant dinner in Rome that he and Gold had with Lemaître in 1952, Lemaître choosing the restaurant himself. Likewise Hoyle, a known anticleric with a hostile attitude toward Christianity, nevertheless enjoyed Lemaître's company and discussing the differences of their theories without any personal rancor getting in the way:

> In 1957 he and I did a 2-week drive together through Italy and the Alps. We only had one disagreement and it wasn't over cosmology. It happened somewhere in Austria, I think in the town of Landeck. It was a Friday, and at dinner that night I ordered a steak while Georges, as befitted his position as a Monsignor of the Church, ordered fish. When the waitress delivered our orders, the steak was good and of a reasonable, sensible size. The fish, on the other hand, was enormous. A prince of a fish, which clearly warranted comment. So in all innocence I said "Now at last, Georges, I see why you are a Catholic." At this, Georges became red-faced

and peevish. For a few seconds I thought the ghost of Martin Luther had tempted me into a serious religio-diplomatic indiscretion. Then in a flash I realized that Georges didn't like fish. He desperately wanted my steak. I would have been happy to swop [*sic*]. But he couldn't, such is life. He had to eat his way through that enormous fish, which like [in] a Germanic fairy story seemed to remain the same size no matter how much of it he consumed.[72]

It was not that Hoyle didn't have any rancor. But he reserved it for the referees and other scientists who often turned down his papers (unjustly he felt) and ultimately forced him to retire early from his chair at Cambridge.

It's fascinating how a certain measure of isolation and alienation seems ultimately to have surrounded all of the major players in the early development of twentieth-century cosmology. Einstein retreated to Princeton and his unified field theory. Eddington died at home during the war, working on his own grand mathematical system. Lemaître was cut off from the rest of the scientific world during the war and in a very real sense never recovered his status. Gamow became increasingly erratic in his work as his drinking became more serious, and Alpher and Herman took jobs outside academia.

For the time being, the 1950s, the steady state theory attracted more attention because it seemed to evade some of the key problems of Friedmann-Lemaître models, the main issues being the timescale problem, the formation of galaxies, and the formation of the heavy elements. Hoyle's expert credentials in the field of element formation due to fusion inside stars also gave him a great deal of credibility. Steady state would remain a more attractive theory, certainly in Britain, until the astronomers in the late 1950s and early 1960s began to discover unmistakable signs of cosmic evolution—the distant galaxies and their compositional difference from closer, younger ones, and the existence of quasars. These discoveries were beginning to undermine the status of the steady state theory even before the discovery of the cosmic background.

9.

THE RETURN OF LEMAÎTRE'S CONSTANT

> The cosmological constant can be compared to the iron rods that are hidden inside the walls of a concrete building. They are no doubt superfluous in a completed construction, but they are indispensable if the structure of today needs to connect with later structures and become one component of a synthesis more vast.
>
> *Georges Lemaître, "Recontres avec A. Einstein"*

IN JULY 1947 Lemaître wrote a letter to Einstein—the last of their correspondence that survives—in which he tried, once again, to persuade the founder of the general theory of relativity to reconsider his dismissal of the cosmological constant.

As has often been written, Einstein considered the introduction of the cosmological constant, the lambda term Λ, as the greatest "blunder" of his life.[73] He had introduced it into the field equations in his 1917 paper, "Cosmological Considerations of the General Theory of Relativity," in order to counterbalance the universal effects of gravity at a cosmic level. Einstein's pure equations suggested a universe that would not be stable. It would either collapse under its own weight—a consequence Einstein had hoped his theory would avoid—or it would expand, something he seems not to have seriously considered, as he knew of no evidence to suggest this at the time. In fact, by 1917 there were already indications from U.S. astronomers—in particular, from the work of American Vesto Slipher at the Lowell Observatory in Flagstaff, Arizona, at the time—that many nebulae exhibited spectra that were redshifted, suggesting movement outward into space. But it was not then known why these redshifts existed or what percentage of the nebulae exhibited them. (Slipher had measured thirteen Doppler-shifted nebulae by 1914, all but two of which were redshifted.) It was not even determined positively that the nebulae existed outside the Milky Way. In short, in 1917 the known universe was a smaller place, and it didn't seem to be changing. In light of this, Einstein assumed as scientists had for so long, that the universe as a whole was static, unchanging. However, his equations were not suggesting any such stability. Einstein realized he needed to maintain a cosmic equilibrium that would support what he considered to be the only cosmic solution to his field equations: a static, spherical universe, finite but unbounded in space, and temporally infinite. He thus introduced Λ, a constant, that at cosmological distances would counteract the force of gravity and maintain the universe in static equilibrium. At smaller scales, that of the solar system, its effects would have to be negligible.

While the lambda term has often been described as ad hoc and arbitrary, Einstein realized that from a mathematical and physical point of view the term did have a natural place in the field equations. In the summer of 1918 he wrote to his old sounding board, Michele Besso, an engineer who had listened to him back in the early months of 1905 when he was working his way to the **special theory of relativity**. And Einstein described to Besso his reasons for adjusting the field equations to allow for Λ:

> Either the universe has a centre, has a vanishing density everywhere, empty at infinity where all the thermal energy is gradually lost as radiation; or, all the points are equivalent on the average, and the mean density is everywhere the same. In either case, one needs a hypothetical constant Λ, which specifies the particular mean density of matter consistent with equilibrium. One perceives at once that the second possibility is more satisfactory, especially since it implies a finite size for the universe. Since the universe is unique, there is no essential difference between

considering Λ as a constant which is peculiar to a law
of nature or as a constant of integration.[74]

These words offer some insight into what specifically was
bothering Einstein about his theory's cosmological consequences: his initial attempt to establish boundary conditions
for a universe that could be *infinite* in space and time simply
didn't work in general relativity.

A cosmos that was infinite in space and time, one dreamed
of by philosophers through the ages, had been what Einstein
preferred. He believed it was necessary. Since Newton's *Prin-
cipia* was published, scientists viewed space and time as infinite, even though this view was not free of contradiction. If
gravitation was truly universal, as Newton posited, then the
universe would collapse under its own weight. As this had
obviously not happened, Newton went on to argue that the
cosmos in space and time must be infinite to avoid this fate.
There were two problems with this argument, however: one
is that making space infinite does *not* rescue the universe from
gravitational collapse, as was shown later by several physicists.
And, as pointed out by Wilhelm Olber with his famous par-
adox and by other astronomers, if the universe was truly infinite in space and time, why was the sky dark at night? There
did not seem to be any natural way out of these contradictory
positions without altering Newton's theory (which some
astronomers did).[75]

Einstein supposed that from the start, his field equations
of general relativity would offer a contradiction-free solution

to this age-old problem. But when he applied the field equations in order to find boundary conditions for infinity (referred to in the first sentence of his letter quoted above), he realized that space, which he believed could not exist without matter in general relativity, would literally fade away. Relying on a familiar and useful statistical argument based on Ludwig Boltzmann's theory of gases, he imagined the finite universe and all the stars as a cloud of gas particles "in equilibrium at some finite temperature. If the number of stars per unit volume was to vanish at the boundary, then . . . it must also vanish in the center of the distribution." This was unlikely. Averaged over the known universe, stellar density was "sensibly constant."[76] So general relativity seemed incompatible with an infinite static universe. Thus, without thinking much more about the consequences, and cognizant of the astronomical observations of the time, Einstein decided to modify his equations to sustain a static universe that was *not* infinite. This was the genesis of the Einstein universe: a closed four-dimensional sphere of stars, finite but unbounded, with its positive curvature determined by the general theory of relativity and balanced in equilibrium by the cosmological constant. The Einstein model was also called the Einstein cylindrical world after representations that dropped two dimensions in order to show a flat two-dimensional world moving forward in time.

As we've seen, Einstein's was not the only cosmic solution to his field equations. Willem de Sitter lobbed his own solution into the ring before the year 1917 was out. De Sitter's was a very fertile solution: a spatially flat universe with nothing in it at all, a model that would later inspire not only Lemaître in his own models but also Fred Hoyle in his steady state theory, more than a decade after de Sitter's death, and Alan Guth with his inflationary theory. Although Einstein disagreed with de Sitter's model, arguing that there could be no space without matter in general relativity, he soon got used to the fact that his own model was not the only possibility: various spin-off models soon followed.

By 1929 the expanding models of Friedmann and Lemaître were given convincing support by Hubble's findings on the galactic redshifts indicating what Hubble cautiously called their "apparent" velocities of recession, which increased with distance out into space. By the early 1930s the larger scientific community concluded that bizarre as it seemed, the universe was expanding like a balloon. Einstein also accepted this conclusion officially by 1931 when he was visiting California. Recall that Friedmann's models did not require a cosmological constant (or rather, the Russian mathematician set the term = 0). As far as Einstein was concerned now, the lambda term, about which he'd always "had a bad conscience" as he told Lemaître, was not only dispensable, he should never have introduced it in the first place.[77] And to Gamow he supposedly called it the "biggest blunder" of his life.

Lemaître and Eddington were not so sure. The cosmological constant was not an arbitrary number, not a number picked out of thin air merely to balance the force of gravity. For them Λ represented something more, something physical.

Lemaître in particular began to treat the constant as the indicator of an actual force with a special role to play in an expanding universe. Unlike Einstein, who used the lambda term on the left side of his field equations (the terms governing the metric tensor, the geometry of space-time), Lemaître switched Λ to the right side, where it acted as a force contributing on a cosmic scale to the energy stress tensor, representing the energy density (and indeed anything: mass, radiation, pressure) that determines the curvature of space-time described on the left side of the equations. Lemaître used Λ as a real negative pressure, a force that was negligible on the small scale of the solar system, but grew proportionally with distance on the cosmic scale.

According to Helge Kragh's *Cosmology and Controversy*:

> In his 1933 address [to the American National Academy of Sciences in November] he suggested an interesting interpretation of the constant, namely, that it may be understood as a negative vacuum density. "Everything happens as though the energy *in vacuo* would be different from zero," he wrote, referring to general relativity applied to regions of space of extremely low density. He then argued that in order to avoid a nonrelativistic vacuum or ether—a medium in which absolute motion is detectable—a negative pressure $p = -\rho c^2$ must be introduced, the vacuum density being related to the cosmological constant by $\rho = \Lambda c^2 / 4\pi G$. This negative pressure is responsible for the exponential (de Sitter) expansion of

Lemaître's universe during its last phase, and it con-
tributes a repulsive cosmic force of $\Lambda c^2 r / 3$.

(Where ρ = density, c = speed of light, Λ = cosmological constant, G = Newton's constant of gravity.)[78]

In this way, Lemaître realized he was able to "play" with various models of expansion, literally speeding up or slowing
down the process at different epochs in cosmic history. This
proved prescient.

Eddington also believed the cosmological constant represented a real force. In fact, he saw Λ as the root cause of expansion itself. Where Lemaître introduced the primeval atom,
what would become the big bang singularity from which the
universe expanded, Eddington always preferred the interme-
diate model, which he and Lemaître had shown could start
expanding from an initial static Einstein state. After a nearly
infinite epoch of stasis, this model began to expand outward,
accelerating under the influence of the cosmological constant.

In his 1922 book, *Space, Time, Matter,* Hermann Weyl also
saw a deep connection between Einstein's gravitation and
electromagnetic theory. For Weyl, who had an early view of
unification between the two forces of gravity and electromagnetism, lambda was a factor that was intrinsically dependent
on the electromagnetic potential, and he wrote, "The cosmological factor which Einstein added to his theory later is part
of ours from the very beginning."[79]

In later years Eddington picked up where Weyl left off.

Although an astronomer by training, he now seemed less interested in lambda's role affecting the rate of cosmic expansion as he was in its role in his fundamental theory (of everything), in particular, lambda's role in connecting gravitation with electromagnetism: "Not only does it unify the gravitational and electromagnetic fields, but it renders the theory of gravitation and its relation to space-time measurement so much more illuminating, and indeed self-evident, that return to the earlier view is unthinkable. I would as soon think of reverting to Newtonian theory as of dropping the cosmical constant."[80]

Eddington died in 1944, but in his last decade he occupied himself, very much as Einstein did, trying to draw up his master theory that would comprehensively describe all nature in purely mathematical terms. He regarded the cosmological constant as one of the absolute constants of nature. Indeed, he regarded the universe as a symphony played on seven "primitive" constants in the same way that music is played on the seven notes of a scale. The seven constants were the mass of the electron, mass of the proton, the charge of the electron, Planck's constant, the speed of light, the constant of gravity, and finally the cosmological constant.[81]

But nothing came of Eddington's theory, and by 1947, when he'd been dead for three years, Lemaître was virtually alone among cosmologists who believed the Λ term had an important role to play in the expansion of the universe. At the cosmic scale, Lemaître argued that Λ did not merely support the universe against collapse, but it also caused the expansion rate *to change with time,* sometimes speeding up, sometimes slowing down the rate of cosmic expansion. Back in the 1930s and 1940s, however, Lemaître had a hard time convincing anyone to take it seriously. And he had no luck with Einstein.

Returning to the subject of their correspondence, Lemaître's 1947 paper, which he sent to Einstein at Princeton, was prepared as part of a collection of essays in Einstein's honor, to be published as part of the Library of Living Philosophers.[82] In his letter he wrote, "I have chosen for subject 'The Cosmological Constant' a subject that I have had sometime the advantage to discuss with you. I remember that the last time I met you at Princeton some of my reasons impressed you some what [*sic*]." Lemaître cited three reasons for retaining Λ in any serious model of the universe. (Typos and misspellings from the following passage have been corrected.)

- that gravitational mass, which has a definite effect, could not have been identified with energy, which is defined but for an additive constant, if theory would not provide some means of adjustment when the zero level of energy is changed at will.
- that the cosmical constant is necessary to get a time-scale of evolution which would definitely clear out from the dangerous limit imposed by the known geologic ages.
- that the instability of equilibrium between gravitational attraction and cosmical repulsion is the only

means to understand an evolution on the stellar scale during the short time available (of some ten times the duration of the geological ages). All that would be impossible without the cosmological constant.

Einstein's reply, dated September 26, 1947, began "I doubt that anybody has so carefully studied the cosmological implications of the theory of relativity as you have." Nevertheless, he went on:

Since I have introduced the term I had always a bad conscience. But at the time I could see no other possibility to deal with the fact of the existence of a finite mean density of matter. I found it very ugly indeed that the field law of gravitation should be composed of two logically independent terms which are connected by addition. *About the justification of such feelings concerning logical simplicity it is difficult to argue. I cannot help to feel it strongly and I am unable to believe such an ugly thing should be realized in nature.* [emphasis added]

Here Einstein's belief in the simplicity and beauty of nature is explicit, a beauty he thought should be manifested in his equations. And there did not seem to be anything beautiful about the balancing factor of the lambda term in his view, although he was not up to the task of trying to justify his feeling in his letter to Lemaître.

He spent the second page of his short letter explaining why he didn't find Lemaître's first point—that Λ allowed gravitational mass to be identified with energy in the general relativity equations—convincing, but he did allow that in regard to the timescale problem in points 2 and 3, "there exists a reason to try bold extrapolations and hypotheses to avoid contradiction with facts. It is true that the introduction of the Λ term offers a possibility, it may even be that it is the right one." This is hardly the flat dismissal of Lemaître's argument portrayed in other accounts. What one wonders here is how much more persuasive Lemaître might have been had he accepted Schrödinger's offer of a visiting professorship at the Institute for Advanced Studies in 1951, where he could have more persistently pestered Einstein in person.

Within a few years of the 1947 correspondence, new measurements of galactic distances made by German-born astronomer Walter Baade at Mount Palomar would force a reappraisal of the Hubble time, giving a cosmic timescale of almost four billion years, much more amenable to the geologic dating of the Earth and theories of solar evolution of that period. (It seems remarkable how far off the mark Hubble was in his initial estimate of the Hubble time and for how long his data was unchallenged.) Hubble's pupil Allan Sandage and the redoubtable Milton Humason would revise the figure even further upward by 1956, prompting Sandage to note "there is no reason to discard exploding world models on the evidence of inadequate time scale alone, because the possible values of H are within the necessary range."[83] *H* here being the Hubble time. In other words, with the reappraisal of the universe's age at upward of four billion years and more, the standard Friedmann-Lemaître models were considered sufficient to explain the cosmic expansion, so there was even less reason to need the cosmological constant.

But Lemaître never let go of it, and his intuition now seems to have been more than justified. Lemaître's use of the constant not only seems to have anticipated the inflationary theory (albeit in a narrow sense) by more than four decades, but it also anticipated observations that by the end of this past century would convince astronomers that the universe's rate of expansion was indeed not constant, that in the past expansion had been slower, and that presently it is faster. Recall that Lemaître's "hesitating" model, one that expands from an initial super-dense state, slows down, and then speeds up in expansion.

Alan Guth, who first posited the inflationary theory of the early universe in 1980, described how Lemaître's work inspired him.

> Although I didn't know it at the time, the exponentially expanding space that I discovered was hardly new—it was in fact one of the earliest known solutions to the equations of general relativity. I had rediscovered the equations of de Sitter's cosmology of 1917, written in a form that was introduced by Georges Lemaître in 1925 as part of his Ph.D. thesis at MIT.[84]

This was Lemaître's "Note on de Sitter's Universe," which showed that the Dutch astronomer's model was in fact the first example, indeed a limited case, of a nonstatic world model.[85]

The inflationary theory of the universe was devised by Guth to deal with a particle physics problem that plagued the standard model of the big bang in the later 1970s. By that time elementary particle theorists were trying to combine the standard model of the big bang with grand unification theories. The problem was that such "GUTs," as they were called, all predicted a surplus of particles called "magnetic monopoles" in the early stage of the universe. A magnetic monopole is a sort of quantum-level magnet that has only one pole (where a standard magnet has a north and south pole of equal strength). According to Guth, if GUTs predicted a surplus of magnetic monopoles, the big bang model then needed to explain what happened to them, since they are not observed anywhere in the cosmos. But Guth also realized his theory dealt with two other nagging problems that plagued the standard model of the big bang.

Since Penzias and Wilson had discovered the cosmic microwave background radiation, it was calculated to have originated about three hundred thousand years after the big bang, when the universe became transparent for the first time,—that is, when photon decoupling occurred, and coalescing matter began to predominate over radiation. Refined measurements of the cosmic background showed it to have the same temperature no matter the direction in space. But calculations also suggested cosmic background radio waves arriving now at the earth from two opposite ends of the cosmos had to have been separated from each other at the three hundred thousand years' initial decoupling point by about one hundred horizon distances. Since there is no way that energy or information can travel more than one horizon distance, one was forced to assume that the universe began in a virtually perfect state of uniformity. This seemed unlikely. It was equally unlikely, cosmologists believed, that the universe just happened to begin with a mass density so close to the critical density needed to keep it balanced (from either expanding forever or collapsing back in on itself).

Astronomers designate Ω (omega) as the ratio of the actual mass density of the universe (as best it can be measured) to the critical mass density. An Ω of 1, for example, would mean the universe is essentially flat, neither expanding forever nor collapsing under its own weight. $\Omega = -1$ would expand forever; $\Omega = +1$ would collapse into a singularity. But why the universe should have started so close to Ω in the first place seemed bizarre. Guth realized that by positing a false vacuum in the early stages of the universe, one exhibiting an outward negative pressure very like that associated with Λ, then one could theorize a sudden, brief but very powerful surge in the expansion of the cosmos—an inflation—that would swell the size of the universe to such an extent in so short a time that the horizon problem and the magnetic monopole problem could effectively be solved. The universe, in effect, could have started in *any* arbitrary mass density and lack of uniformity and still have evolved into what was measured today. In the same way, a deflated balloon with lots of wrinkles and folds rapidly loses all of its defects when inflated rapidly to a substantial size. Because of inflation, Guth's theory also suggests that the observable universe may be just a sliver of its actual size.

Lemaître seems also to have anticipated this. In 1933 and 1934 he and Tolman wrote papers exploring nonhomogeneous solutions to the field equations, arguing that parts of the universe could be expanding while others were contracting. As Tolman wrote, "regions of the universe beyond the range of our present telescopes might be contracting rather than expanding and contain matter with a density and stage of evolutionary development quite different from those with which we are familiar."[86] Much the same has been said of the vast universe described by the inflationary model, that the current observable universe is but one bubble in a sea of expanding and contracting pocket universes. While not everyone accepts the inflation theory, cosmologists generally utilize it as a key part of current big bang models.

Is it possible the cosmological constant itself could be responsible for the false vacuum behind the inflation? Guth did not believe so (at the time of his book), because lambda was assumed to be constant in most relativistic models.

Directly or in combination with another source of energy, it is clear now that Λ is deeply involved in what has turned out to be the most significant discovery of the last half-century— that the universe is accelerating. What has excited astronomers and cosmologists over the last seven years is the realization that somehow the outward pressure force is still at work in this expansion phase.[87] Whether it is Λ or Λ in combination with dark energy of unknown origin remains to be seen.

When a star explodes, astronomers have a rare opportunity to see what elements are produced in the calamity. The last report of a supernova in our Milky Way galaxy was made by Johannes Kepler in 1604. It is believed that they occur much more frequently than that, but our position in the Milky Way obscures Earth from getting a better sample. In 1987 a star exploded in the Milky Way's neighboring satellite galaxy, the Larger Magellanic Cloud, giving astronomers a bounty of new observations and spectra. As data has grown over the years, so have the classifications of supernova types.

The standard supernova, meaning the one that's least complicated to explain, is that of a massive star eight or more times the mass of our sun. It explodes when the star has exhausted all of its hydrogen and helium, and has even fused its carbon to the point where there is nothing left to burn in order to counteract the force of gravity; the star lights up in the process of collapse. This is a **Type II supernova** (SN II) in the astronomer's lexicon, and usually presages the creation of a neutron star or black hole in its place.

Impressive as SN IIs are, there is another wildly spectacular SN type that has become crucial to measurements of astronomical distances: the Type SN Ia. A **Type Ia supernova** is an already shrunken star, a white dwarf, that accretes more mass by sucking matter off of a nearby companion star until it reaches a limit (1.44 solar masses defined by the astronomer Subrahmanyan Chandrasekhar). At this point the star explodes in an eruption whose brightness rivals the brightness of entire galaxies. It also explodes in a way that can be used as a measuring standard. If one knows the absolute luminosity of the explosion type, one can compare the apparent luminosity of the Type Ia wherever it occurs in combination with its redshift to get a much better approximation of distance than the older Cepheid variable standard used by Hubble and Humason back in the 1920s and 1930s.

Starting in the late 1980s, Saul Perlmutter and his team at Lawrence Berkeley National Lab in San Francisco began sampling large patches of galaxies in order to collect data on as many Type Ia's as they could find. Using the new data from Type Ia's, Perlmutter was expecting to get a new, more accurate estimate of how the universe's expansion was slowing. What he discovered was the exact opposite. By 1998 he realized that the farther out into space he measured, the slower the apparent velocities of the galaxies appeared. Galaxies that were closer were receding faster than galaxies that were farther away. Even after taking into account how much longer ago the light traveled from the more distant galaxies (up to twelve billion light-years), it was impossible to avoid the conclusion that the universe was expanding faster *now* than it was billions of years ago. As startling as this data was, it was confirmed independently when a team led by Harvard astrophysicist Robert Kirshner found the same results in his collected samples.

Almost overnight, an entire generation of cosmologists found themselves revisiting Einstein's equations and realizing that a certain term needed to be reinserted: the cosmological constant. This last decade of discovery has been in a broad sense a vindication of Lemaître's open-mindedness about Λ as an intrinsic part of the Einstein field equations. The new assessment of cosmic expansion cries out for an explanation and is one of the chief subjects of interest in current cosmological theory. The cosmological constant, so long confined to the dustbin of twentieth-century cosmology, is being examined again as a likely candidate to explain why the universe is accelerating. In the end, Einstein's "blunder," if he ever called it that, seems not to have been a blunder at all, but another clue in the long search for the ultimate picture of the universe.

The issue is not cut and dry, of course. Most cosmologists believe it is unlikely, based on current theory, that the Λ constant alone can explain the acceleration. In the field equations of general relativity, the lambda term represents a value that is close to zero, and must be close to zero without causing the theory to describe a universe that would blow apart. Quantum theory, however, gives a different picture of what the cosmological constant should be. Elementary particle theorists believe that Λ is a measure of the energy density of the vacuum in space, something Lemaître suggested himself back in 1933. But the energy density in quantum terms is expected to be large—in fact, it is about 120 orders of magnitude larger than the classic close-to-zero values used in the field equations of general relativity. So the discrepancy between general relativity and quantum mechanics on this matter is huge. No one knows how soon it will be solved. But most astronomers and physicists agree that it will remain unsolvable until a new quantum theory of gravity is developed that can unify general relativity with quantum mechanics.

10.

SEEING THROUGH THE SINGULARITY

Lemaître had a combination of talents that was fairly rare in general relativity at the time. He had an excellent background in mathematics, especially in differential geometry, and a strong physical intuition.

Jean Eisenstaedt, "Lemaître and the Schwarzschild Solution"

IN A SENSE, Lemaître's 1925 "Note on de Sitter's Universe," which was part of his MIT PhD thesis, was the foundation for all of his subsequent work in cosmology. Not only were the seeds of his 1927 paper on the expanding universe planted there, but also sown was the approach that would inspire an often overlooked 1932 paper that would change the way cosmologists regarded the Schwarzschild solution, effectively paving the way for the theoretical development and modeling of the grandest cosmic artifacts uncovered by general relativity: black holes.

In his "Note" Lemaître showed that through a poor choice of coordinates, de Sitter found only an apparently "static" solution of Einstein's equations, one showing an empty universe that was spatially flat. But if one applied de Sitter's solution over any arbitrary set of coordinates, it revealed a model of the universe that was not stationary. Particles introduced into it would drift apart, with redshifts being the clue to their velocities, and Lemaître suggested at the time that actual redshifts of extragalactic nebulae would be a confirmation of the model. It's important to note that while de Sitter's model was later realized to be the first limited case of an expanding space model, the redshift of objects in de Sitter's space was not interpreted at the time as due to the expansion of space itself.

Just as Lemaître thought the problem with de Sitter's model was a poor choice of coordinates, he also suspected that Karl Schwarzschild's first complete solution to Einstein's equations, for a perfect sphere of fluid matter, was also only a limited case of a more general, dynamic solution. In his 1916 paper Schwarzschild effectively set a limit on the radius of a sphere in gravitational contraction.[88] The famous Schwarzschild radius, $2Gm/c^2$ (where G = Newton's gravitational constant, m = the mass of the sphere and c = the speed of light) is known as the limit a star can shrink to before it collapses into a black hole. This is the modern interpretation, but it is *not* how the radius was viewed at the time Schwarz-schild derived it, either by its author or by Einstein and his colleagues.

Schwarzschild found two solutions for the sphere in his paper—an interior solution and an exterior solution:

> In the interior Schwarzschild solution, matter is described by a fluid sphere of constant density and radius r = a (Schwarzschild 1916). Schwarzschild noticed that, in his model, the pressure becomes infinite at the center as soon as the radius of the sphere is equal to the Schwarzschild limit. Since the radius $2Gm/c^2$ of the Schwarzschild singularity is smaller than the Schwarzschild limit, the latter could be used to lay worries about the former to rest. The Schwarzschild singularity, it seems, will be physically inaccessible.[89]

In other words, because the pressure of the sphere for the interior solution became infinite in his application of the equations, Schwarzschild believed that "Thus, there is a limit concentration above which an incompressible fluid sphere cannot exist." The *$2Gm/c^2$* Schwarzschild limit was deemed the end of the matter. In essence, Schwarzschild's conclusion allowed Einstein and his contemporaries to argue that no physical state existed beyond this horizon.

Eddington called the Schwarzschild limit the "magic circle," on which light and particles would congregate but never penetrate. Today, of course, this limit is described as something else, the "event horizon," the limit beyond which a star must shrink before it collapses into a black hole, and Schwarzschild is by association often credited with the "discovery" of black holes in theory as a prediction of general relativity. But in fact Schwarzschild regarded the limit itself as the singularity, one beyond which, as Einstein believed, nothing was *physically realized in nature.* But Schwarzschild's limit is not really the limit that later astrophysicists regard as the actual singularity, the signature of a black hole. It is considered the event horizon beyond which matter and energy cannot escape from the black hole within. What happened to change the view of Scharzschild's solution?

In 1932 Lemaître showed that the real singularity is in fact the point at which the radius of the sphere collapses beyond $2Gm/c^2$ to zero, and he devised a new solution to Einstein's field equations to show how such a singularity could be revealed. Starting from Schwarzschild's interior solution, Lemaître argued that by assuming a sphere of zero pressure—one of dust rather than fluid—the apparent "singularity" of Schwarzschild's solution was just that: apparent, a fiction. One could push further inward than the alleged limits of the "magic circle." One could in fact push right to the point where the radius = 0. Why? Because, Lemaître remembered, the equations of the Friedmann universe allowed "solutions in which the radius of the universe goes to zero. This contradicts the generally accepted result that a given mass cannot have a radius smaller than $2Gm/c^2$." (Recall that Friedmann's papers were written in 1922 and 1924, six and eight years after Schwarzschild's untimely death.)

What fascinated Lemaître was exploring the status of the space between a collapsing interior sphere and Schwarzschild's classic singularity, between the space where $r = 0$ and $r = 2Gm/c^2$. Schwarzschild had used two coordinate systems to examine each. Lemaître used only one to examine both. Mathematically, what Lemaître discovered was "a form of the Schwarzschild line element [meaning, the squared distance between two close points of space-time] that explicitly shows that the only singularity of the solution is at $r = 0$. There is no singularity at $r = 2Gm/c^2$."[90]

Although Lemaître's 1932 paper, "The Universe in Expansion," did not seem to arouse much interest from the larger community at the time (it was never translated into English during his lifetime),[91] it's clear that Richard Tolman, with whom Lemaître worked for two months during his stay at Cal Tech that year, accepted the Belgian physicist's **dust solution** and used it in a subsequent paper of his own in 1934.[92] Hermann Bondi, cofounder of the steady state theory, also used it in a paper a few years later, and for this reason it has often been known as the Tolman-Bondi solution.[93]

In 1934, taking his cue from Tolman's paper, Irish physicist John Lighton Synge (nephew of the great Irish playwright John Millington Synge) also used Lemaître's dust solution to show that a pressure-free cloud of particles could contract further than the Schwarzschild singularity.[94] But he did not credit Lemaître with the dust solution until a similar paper he

Lemaître celebrating his 70th birthday with colleagues Odon Godart and Andre
Deprit. Image courtesy of Archives Lemaître, Institut d'Astronomie et de
Geophysique Georges Lemaître, Catholic University Louvain.

wrote over fifteen years later. Synge's 1934 paper, a definite
forerunner of J. Robert Oppenheimer's work, seems to have
aroused as little reaction at the time as Lemaître's initial foray
into the subject two years before. A widely respected physicist, Synge was born just three years after Lemaître in 1897,
but he lived until 1995.

It was the work of Tolman whose adoption of Lemaître's
solution directly inspired J. Robert Oppenheimer and Hartland
Snyder's classic paper "On Continued Gravitational Contraction," in 1939.[95] The year before, Oppenheimer had written to
Tolman for ideas and feedback for a paper he and George
Michael Volkoff were writing on the subject of neutron stars.[96]
So they were in touch throughout the process that led to the
seminal paper on collapsed stars (they were not called "black
holes" until John Archibald Wheeler coined the term in 1968).

Oppenheimer and Snyder's is the first paper to explicitly use the field equations with Lemaître's and Schwarzschild's solutions to describe the physical collapse of a star—not just an idealized mathematical sphere or cloud of particles. The introductory abstract to their paper famously states:

> When all thermonuclear sources of energy are exhausted a sufficiently heavy star will collapse . . . the radius of the star approaches asymptotically its gravitational radius; light from the surface of the star is progressively reddened, and can escape over a progressively narrower range of angles . . . an external observer sees the star asymptotically shrinking to its gravitational radius.

The importance of Lemaître's contribution cannot be overemphasized. "Lemaître's solution is not only one of the first general dynamical solutions with spherical symmetry in general relativity, it also allows us to describe the complete evolution of a star, its interior as well as [its] exterior gravitational field, in a *single* coordinate system" (emphasis original).[97]

Somehow, between Tolman and Oppenheimer, the credit for Lemaître's dust solution to the field equations was lost. There's nothing suspicious or malicious about this, to be sure. Tolman, after all, learned the solution from Lemaître directly at Cal Tech, after Lemaître had submitted his paper to *Publication de Laboratoire*. But as noted above, the paper was not translated into English. Tolman nevertheless did credit Lemaître in his 1934 paper. Oppenheimer in turn credited Tolman in his 1939 paper, not acknowledging Tolman's dependence on Lemaître. This happens frequently in scientific papers. And as previously mentioned, after Bondi's paper in 1947, many believed the dust solution was actually due to him. To further complicate matters, according to Eisenstaedt's account, Charles Misner's later standard text on general relativity erroneously stated that Oppenheimer and Snyder's model was homogeneous, when it wasn't;[98] true to the method Lemaître devised, Oppenheimer and Snyder made their model pressure-free everywhere, but the density was a function of radius r and time t in accordance with Lemaître's dust solution. The point is that Lemaître's insights were key to the development of black hole theory. Indeed, "he understood the inevitability of collapse to zero volume and the fictitious character of the Schwarzschild singularity, insights that even Oppenheimer and Snyder failed to reach."[99]

To step back at this point is to realize that Lemaître's insights were in fact key in almost all the important milestones of early modern cosmology, from the expanding universe, to the cosmological constant, to black holes. He was first to see how the Einstein and de Sitter models were but two limited cases of a larger body of expanding universe models; he was the first to see that such models had to evolve from a super-dense state; and perhaps most importantly—from the very beginning—he was the first to tie the predictions of relativity about cosmology to *actual astronomical observations.* How did he do it? What advantage did Lemaître have over his contemporaries—even Einstein—that led him to fully imagine all these fruitful cosmological consequences of the general theory of relativity, consequences that even its author could not or would not entertain?

In a contribution to the *Einstein Studies,* Volume 5, Professor Jean Eisenstaedt offered an insightful assessment, one worth quoting in full:

> One of the most important characteristics of Lemaître's approach, I think, is the subtle interplay between local and global concerns in his work; the stars and the cosmos, contracting nebulae and the expanding universe, the condensation of a star and the collapse of the universe. In a way, Lemaître was able to describe the local in the global: a star is embedded in the universe, and the Schwarzschild solution is described in the same coordinates as Friedmann's solution. This tendency to combine the local and the global, the awareness of the parallels between cosmology and the treatment of an individual star, enabled Lemaître to view things in new and unexpected ways, to look, so to speak, at the Schwarzschild singularity from the interior, or at the universe from the exterior. It was this general approach and his extraordinary facility in delicately manipulating the equations of the universe that enabled

Lemaître to shake off the dogma of the impenetrability
of the Schwarzschild singularity.[100]

Moreover, Lemaître's comparative youth almost certainly
helped him here as well. In a sense, he was the first cosmologist to grow up with Einstein's physics rather than Newton's.
His contemporaries—in many ways mentors—such as Einstein, Eddington, de Sitter, even Tolman and Robertson were
older, more classically trained physicists than Lemaître was.
And they, like Einstein, tended to approach the problems of
the new cosmology in neo-Newtonian terms, "using the
methods of post-Newtonian approximation, constructing
and endorsing a *neo-Newtonian* interpretation of general relativity."[101] Lemaître was free of this approach, free to play
with the relativistic field equations in a way that alerted him
to possibilities that were not obvious to his mentors.
Remember, too, that by training Lemaître came to the
physics of relativity directly from mathematics—rather than
the other way around, which was the common course for
most physics students: to learn the math as the physics of a
system of thought required it.

But there was another factor as well, one that perhaps at
first is not obvious: the influence of Lemaître's vocation; "as a
priest he probably felt a closeness to God that may have given
him a feeling of freedom in front of Creation . . . Lemaître
aimed at combining the global and the local: is there a question more suitable for a priest? To be sure, his answers, it
seems, were strictly physical and mathematical . . . Lemaître

cannot be accused of confusing science and religion."[102] This brings up, finally, the matter of Lemaître's faith and his vocation: how did they influence his outlook on science, on relativity and cosmology, and how did his science influence his faith?

11.

CATHEDRALS IN SPACE

When I was talking with Lemaître about this subject and feeling stimulated by the grandeur of the picture that he has given us, I told him that I thought cosmology was the branch of science that lies closest to religion. However Lemaître did not agree with me. After thinking it over he suggested psychology as lying closest to religion.

Paul Dirac, "The Scientific Work of Georges Lemaître"

Hence that the world began to exist is an object of faith, but not of demonstration or science.

St. Thomas Aquinas, Summa Theologiae,
Question 46, Article 2

> The whole question whether the world had a beginning or
> not is, in the last resort, profoundly unimportant for theology.
>
> *E. L. Mascall,* Christian Theology and Natural Science

[The title for this chapter was inspired by a long forgotten essay on science fiction by James Blish, writing under the pen name of William Atheling Jr., on the subject of religion and how it was treated in futuristic stories and novels of the 1940s and 50s. Blish was a constant critic of the genre, himself the author of a celebrated novel about a Jesuit biologist who discovers "an impossible world," one, in sharp contrast to Earth, without any gaps in the fossil record, an apparently perfect case study of Darwin's theory at its purest.]

ERNAN MCMULLIN WAS a graduate student in physics in 1951. Like many of Lemaître's pupils and colleagues, he found the monsignor a man of almost irrepressible enthusiasm and constant good cheer. McMullin and Lemaître were attending a graduate seminar together at Louvain in the fall of that year. But one day in late November, after Lemaître had been away to Rome and a meeting of the Pontifical Academy of Sciences, McMullin later wrote that he "could recall very vividly, Lemaître storming into class on his return from the Academy meeting in Rome, his usual jocularity entirely missing. He was emphatic in his insistence that the Big Bang model was still very tentative, and further that one could not exclude the possibility of a previous cosmic stage of construction. Lemaître

Lemaître meets Pope Pius XII, 1939, after becoming a member of the
Pontifical Academy of Sciences. Photo courtesy of Archives Lemaître, Institut
d'Astronomie et de Geophysique Georges Lemaître, Catholic University Louvain.

was not mentioned in the pope's speech, though a member of
the Academy."[103]

What had happened to visibly sour Lemaître's mood? His
own pope, in essence, had just stepped over the line, publicly
expressing the view that Lemaître's expanding model of the
universe, his primeval atom theory, offered virtual proof that
the creation story in the Book of Genesis was now substanti-
ated by science. This drove Lemaître crazy, and from this
point to the end of his life, he felt the incident—a true
gaffe—had confirmed the suspicions of many scientific colleagues (especially Fred Hoyle, for example, and William
Bonnor) that the big bang theory was justifiably suspect
because Lemaître's faith rather than physics had inspired the

theory of the expansion of the universe from its origin in a super-dense state.

Pope Pius XII has been the subject of great controversy over the decades since the end of World War II, due to the question of his role in the Church's policy toward the Jews at the time of the Holocaust. The son of Italian aristocrats with a long family history of involvement in Vatican government, Eugenio Pacelli (the future pope's birth name) was highly educated and had broad interests. But he was also a classic Church bureaucrat, a member of the Vatican Curia who was groomed throughout his career to be pope one day. That day came in 1939, after he had acted as Papal Nuncio to Germany for the years leading up to the period just prior to the war, when Pope Pius XI died.

There have been a number of recent biographies of the man, almost none of them entirely objective—and some whose historical veracity has been seriously questioned by historians.[104] This is not the place to evaluate the pontificate of the wartime pope, but what comes across in all of the biographies, whether pro or con, is Pius's intense devotion to the preservation of the institution of the Church; indeed it seems to be the locus around which all of his momentous and tragic decisions were made.

Paul Johnson paints a portrait of Pius as a tireless but eccentric enthusiast for all aspects of modern life, often spending hours in isolation, reading articles on science and technology, because he felt there was no aspect of the modern world the Church should not concern itself with.[105] He issued guidelines, for example, on motion pictures, radio, and television; on how Catholics in the profession should use the media for instructive purposes; and how Catholics in general should digest the media. Reading these now in the days of "reality" TV and of war footage that includes broadcasted beheadings and executions of civilians by religious fanatics, the pope's enthusiasm for the media as a tool for the good of society seems touching rather than naive.

Given the highly publicized and rapid growth of relativistic models of cosmology, it seems Pius could not resist making some kind of statement about the exciting new field, and he found an opportunity to do so in November of 1951. But Lemaître, in spite of being a key figure in the Pontifical Academy since its inception, had not the slightest idea it was coming. Pius XII's predecessor, Pius XI, had personally appointed Lemaître to membership in the Pontifical Academy of Sciences in 1936 when the academy was created. And John XXIII appointed Lemaître its president in 1960 at the outset of the Second Vatican Council. The academy was a truly open organization, with members (limited to seventy) gathered from all walks of life, all fields of science, and all religious persuasions (including those who had none), to broaden the Church's awareness of the latest trends in science and to report to the pope on an annual basis. Meetings, called Study Weeks, could take place almost monthly during the year.

That November of 1951 Lemaître had been invited to Rome at the end of a Study Week for a Solemn Audience. There was no inkling that the pope was preparing to address the issue of

cosmology and its relation to the faith, so Lemaître was broad-
sided on November 22 when Pius stated to the academy, several
cardinals, and the Italian minister of education:

> What was the nature and condition of the first matter
> of the universe? The answers given differ considerably
> from one another according to the theories on which
> they are based. Yet there is a certain amount of agree-
> ment. It is agreed that the density, pressure and tem-
> perature of primitive matter must each have touched
> prodigious values.
>
> Clearly and critically, as when it [the enlightened
> mind] examines facts and passes judgment on them, it
> perceives the work of creative omnipotence and recog-
> nizes that its power, set in motion by the mighty *Fiat*
> of the Creating Spirit billions of years ago, called into
> existence with gesture of generous love and spread over
> the universe matter bursting with energy. Indeed, it
> would seem that present-day science, with one sweep
> back across the centuries, has succeeded in bearing wit-
> ness to the august instant of the *Fiat Lux,* when, along
> with matter, there burst forth from nothing a sea of
> light and radiation, and the elements split and churned
> and formed into millions of galaxies.106

Needless to say, the pope's statement made headlines. The
December 3 issue of *Time* magazine, for example, was titled:
"Behind Every Door: God."107

One physicist, predictably, who seems to have found the entire controversy an endless source of amusement, was George Gamow. Not only did he take a chunk of the pope's statement and append it to the introduction of a paper he wrote a year later, raising a few eyebrows as he hoped,[108] but he also apparently wanted to continue encouraging the pope's incursions into the realm of cosmology and religion, by feeding him articles via an archbishop he knew would deliver his material directly to the Vatican doorstep. "He never had anything to do with the English article 'the,'" Hoyle later remarked of Gamow, "and he went through a phase when he was forever quoting the pope: 'Pope say this . . . ,' or 'About mangan (manganese) pope says that . . .'"[109]

Lemaître was clearly upset. Having been appointed to membership in the Pontifical Academy by Pius's predecessor because of his standing among cosmologists, he was understandably dismayed as to why he had not at least been consulted about the pope's address in advance. Within a few months, both Lemaître and the director of the Vatican Observatory, the Jesuit astronomer Daniel O'Connell, met with the pope to explain that such blatant connections drawn between science and theology would not help the cause of the Church nor the progress of science. And less than a year later, when the pope addressed a gathering of 650 astronomers at Castel Gandolfo, this time he refrained from discussing the religious and metaphysical implications of the big bang theory. To that end, Lemaître and O'Connell's intervention seems to have succeeded. But the damage had been done, as far as Lemaître could see. At this point in his life (he was fifty-seven years old now), if he felt less drawn to developing his cosmological theory any further because of his increasing interest in numerical computing, the embarrassment of having his cosmological work publicly held up as a proof of creation discouraged him even more.

The truth is that by this time Lemaître had not publicly changed or updated his theory from the form he had delivered in 1931. (There are at least two unpublished papers he left at the time of his death, however, that show he had been thinking a lot more about quantum theory and its role.) He had made no contact or collaboration with Gamow to tie his theory into the very progress of elementary particle and quantum physics he himself had said needed to take place in order for the next stage of his theory to proceed. Perhaps Gamow's comic reaction to the pope's address may explain why Lemaître's colleague during those years, Odon Godart, could not interest Lemaître in contacting Gamow about collaborating on the big bang theory.

As much as they outwardly differed on the subject, it's clear that Lemaître and Fred Hoyle had a great deal of professional respect for each other's work. They were both outstanding mathematicians in addition to being physicists. It's possible that Lemaître, like many physicists at the time, found Gamow a bit over the top, precisely because his tendency to engage in pranks undercut the seriousness of his work. His antics following the pope's address may have confirmed that suspicion for Lemaître. Coupled with Gamow's legendary shoddiness in mathematics (and his drinking habits), the imaginative Ukrainian's propensity for pranks might have been a major disincentive for the Belgian physicist. Indeed, Gamow was already burning bridges that eventually led to his later isolation in the 1960s. He would die in 1968 of cirrhosis of the liver, a truly tragic figure, given his talents and the inspiration he had once been able to engender in researchers and colleagues. In any case, Lemaître dismissed the idea of collaborating with him, and in hindsight, this may be viewed as a mistake. The fog of missed opportunity and confusion surrounding the "rediscovery" of Gamow's theory by Peebles and Dicke in 1964, just prior to Penzias and Wilson's discovery of the cosmic microwave background radiation, might have been avoided had Lemaître and Gamow teamed or at least consulted to keep the theory in circulation during the 1950s. The background radiation might have been discovered as much as a decade earlier than it was.

Clearly, Lemaître felt his theory had been undercut by the pope's interference. And he did not pursue further elaboration of it. When he attended the 1958 Solvay conference, he felt uneasy—and self-conscious. Odon Godart recalls that when Lemaître entered the hall for a lecture at a later conference in California is 1961, Hoyle, who was already seated, alerted the man sitting next to him with the mocking quip, "Look, this is the big bang man."[110] Hoyle might well have felt reason to be glib. The steady state theory seemed to many at Solvay as viable a theory—if not more so—than Lemaître's unfortunately named primeval atom. And Hoyle's theory did not suffer the seemingly embarrassing implication of a Creator.

The debate or clash between religion and science has been going on since the middle of the nineteenth century, when Darwin's theory of evolution suggested origins to the human race that were undeniably disturbing even to complacent Victorians used to digesting their weekly sermons with a cup of tea after the local vicar had done his duty for the Sabbath. One could—and journalists often do—set the debate as far back as the Galileo affair, but that tragic misunderstanding between Pope Urban VIII and Galileo, which provided fodder for dramatists with an axe to grind, such as Bertolt Brecht, was more the result of an avoidable clash of

Lemaître in one of the last pictures taken of him, 1962, after dinner with Godart and Ms. Martholome at Louvain. Image courtesy of Archives Lemaître, Institut d'Astronomie et de Geophysique Georges Lemaître, Catholic University Louvain.

egos rather than an unavoidable conflict between the increasingly confident realm of natural science and the supposedly nervous domain of religion. The Jesuits, for example, were already teaching Copernican astronomy at the time Galileo began reporting his telescopic observations, and they were not happy when the fracas encouraged the Vatican to suppress the Polish cleric's classic book *De Revolutionibus Orboeum Caelestium.* It's no little irony that not only was Galileo unable to actually prove the earth revolved around the sun, but also that for purely personal reasons he dismissed Kepler's historic breakthrough on the elliptical orbits of the planets when the latter wrote to him, thus forfeiting a superb chance for the Italian astronomer to defend himself with a sound mathematical system of planetary motion that *agreed with astronomical observations.* Newton would later derive Kepler's three laws of motion from his own law of universal gravitation. Galileo would spend the rest of his life under ignominious house arrest for nothing.

Beginning in the late nineteenth and early twentieth centuries, with the continued progress of physics and archaeology and paleontology, more scientists and philosophers excited by scientific discoveries dismissed organized religion because of a growing sense that it fostered ignorance about the natural world and spread superstition. The Bible was often considered the chief culprit. This is an oversimplified view, to be sure, but certainly Edwin Hubble, Werner Heisenberg, and Bertrand Russell, to name a few, were more typical of scientists with a skeptical attitude toward religion than many of the scientists of the century preceding them. With the exception of Charles Darwin, up until Einstein most of the great breakthroughs in science were achieved by religious men—meaning scientists who were either practicing Christians or deeply influenced by their faith: James Clerk Maxwell, Gregor Mendel, Max Planck.

Einstein was famously uninterested in the workings of organized religion, but it's impossible to read his statements about his own work without finding constant metaphysical references to "Der Alter," the Old One, as he called God. At the same time, Einstein could be startling in his observations about religion, as when he told George Sylvester Viereck of the *Saturday Evening Post* in 1929: "As a child, I received instruction both in the Bible and in the Talmud. I am a Jew, but I am enthralled by the luminous figure of the Nazarene . . . No one can read the Gospels without feeling the actual presence of Jesus. His personality pulsates in every word. No myth is filled with such life."[111]

Einstein often pointed out that relativity theory was itself rooted in a deep-seated belief—indeed what might be called a stubborn article of faith with Einstein—that the universe worked on basically simple universal principles. The entire edifice of relativity started out as a dogged determination to find a deep symmetry between Maxwell's equations of electrodynamics and Newton's system of mechanics. In essence, it was inconceivable to the young Einstein of 1905 that the two systems should be separated by accidents of their mathematical and theoretical development. So he set about the business of finding the asymmetries between the two systems (which turned out to be Newton's assertions of absolute space and time) and eliminated them.

In any case, by the time Lemaître had come to the fore with his expanding model of the universe, the oppositions between science and religion had become fodder for journalists. And it's interesting, for this reason, to go back to the early interviews with Lemaître when he returned to California in 1933 and how he responded to the already typical question—just how did a priest-physicist reconcile his faith with his science?

> The writers of the Bible were illuminated more or less—some more than others—on the question of salvation. On other questions they were as wise or as ignorant as their generation. Hence it is utterly unimportant that errors of historic or scientific fact should be found in the Bible, especially if errors relate to events that were not directly observed by those who wrote about them.
>
> The idea that because they were right in their doctrine of immortality and salvation they must also be right on all other subjects is simply the fallacy of people who have an incomplete understanding of why the Bible was given to us at all.[112]

You can already detect a note of impatience with Lemaître's answer, as in, *Why should this be a question?* But it was.

Cosmology, like evolution, brought the issue of science

and faith to the forefront as soon as Lemaître detailed his expansion model and its origin with time $t = 0$. If there was a temporal origin to the universe's evolution, then it seemed to many scientists to imply the act of creation. In Lemaître's view it did not have to imply that. In many ways, he felt, the argument is based on a misunderstanding of terms—one that many scientists are prone to make and one that theologians are less likely to: that assumption is, what a theologian or philosopher means by *creation* is the same thing as what a physicist means by *origin*. Lemaître indeed would not even have allowed a term like "creation" to be used credibly in a scientific paper. By its very nature, the word describes something that is by definition scientifically unverifiable—how in principle could any experiment or theoretical quantification be made of an act or process (for want of a better term) that by definition precedes all things, including time, space, and matter? Lemaître never made this mistake. But Pope Pius XII did—and so did Fred Hoyle.

Hoyle was a perfect example of the suspiciousness of all things metaphysical (and anti-Catholicism) prevalent among some scientists at the time. "Both Catholics and Communists argue by dogma," he once wrote. "An argument is judged 'right' by these people because they judge it to be based on 'right' premises, not because it leads to results that accord with the facts. Indeed, if the facts should disagree with the dogma then so much the worse for the facts."[113] And although Hoyle got along well with Lemaître, he may have made an exception for the Belgian's collar, for his anticlericalism was as vehement as his hostility to organized religion. He once opined, in the middle of his popular science book *Ten Faces of the Universe,* that the problems of Northern Ireland could be settled by jailing all the clergy on both sides.[114]

No friend of the big bang theory, precisely because of what he regarded as its metaphysical overtones, Hoyle was surprised to see himself taken to task by Soviet physicists who objected to his use of the word "creation" to describe the development of free hydrogen in interstellar space to buttress his steady state theory. "Judge my astonishment on my first visit to the Soviet Union when I was told in all seriousness by Russian scientists that my ideas would have been more acceptable in Russia if a different form of words had been used. The words 'origin' or 'matter-forming' would be O.K., but creation in the Soviet Union was definitely out."[115] But the Russians were quite right. The terminology *does* make a difference and a distinction, and the failure to make the distinction has caused (and continues to cause) confusion and misunderstanding between scientists and theologians (and the general reading public).

Christian theology is unique in that for the first time it defined the world's existence as dependent on an act of creation *out of nothing.* Scientists always presuppose some form of quantifiable matter or substance from which systems can evolve and be subject to experiment. It frankly doesn't go into the cause of the being of the initial conditions, only the state of the initial conditions. After all, Hoyle's steady state theory required—just as much as Lemaître's theory—a coming into being out of nothing, in this case the constant creation of atoms of hydrogen in empty space. The reason Hoyle, Bondi, and Gold's theory was considered less offensive, metaphysically speaking, is that the microscopically small creations of hydrogen atoms took place in an already existing cosmos. But Lemaître argued that the temporal origin of his cosmological model in no way necessitated the conclusion that it was the moment of divine action from outside. This is what got him so angry about the pope's statement. His primeval nucleus could just as easily be the beginning of a new phase of evolution of an eternal universe that went through oscillations, cycles of expansion, and collapse before the big bang took place. (This, for example, is what Gamow believed when he argued with Hoyle about the theory.) It did not necessarily imply an absolute creation out of nothing. Nevertheless, many scientists admittedly resisted the big bang theory because for them it seemed to imply the moment of creation.

English physicist William Bonnor clearly believed this and said so in his book, *The Mystery of the Expanding Universe*.[116] Hoyle, of course, made the term "big bang" famous—but only by deriding the concept. Even Howard Robertson, with whom Lemaître worked, did not like the implication of the super-dense nuclear state from which all things evolved. Lack of religious persuasion was not a factor, either. Arthur Stanley Eddington, a devout Quaker who risked imprisonment for his pacifism during World War I, and probably the most religiously inclined specialist in relativity and cosmology besides

Lemaître, made no effort to conceal his "repugnance" for a conceived temporal beginning of the universe.

The associations with creation did not really undercut the power of relativistic cosmology. Even Hubble was willing to consider Lemaître's a viable model. On the other hand, more overtly Christian-inspired cosmologies, such as the model based on the kinematic relativity of Edward Arthur Milne in the 1940s, barely caused a stir among scientists, mostly because of what were regarded as mathematical weaknesses. And William H. McCrea, a distinguished physicist who supported steady state cosmology in the 1950s, did so—to the evident distress of Hoyle, Bondi, and Gold—out of blatant Christian conviction.

So the issue is by no means a simple one. For Lemaître, religion and science were two utterly different paths to truth, and neither should interfere with the other. Whether or not Lemaître felt slighted by Hoyle's jibe at the 1961 California conference, he had a response to those who believed the big bang theory was inspired by religion:

> As far as I can see, such a theory remains entirely outside any metaphysical or religious question. It leaves the materialist free to deny any transcendental Being. He may keep, for the bottom of space-time, the same attitude of mind he has been able to adopt for events occurring in non-singular places in space-time. For the believer, it removes any attempt at familiarity with God, as were Laplace's chiquenaude [or flick of God's

finger] or Jean's finger [of God]. It is consonant with
the wording of Isaias speaking of the "Hidden God,"
hidden even in the beginning of creation. . . . Science
has not to surrender in face of the Universe and when
Pascal tries to infer the existence of God from the supposed infinitude of Nature, we may think that he is
looking in the wrong direction.[117]

Any cursory perusal of newspaper articles on science and religion today, over seventy years after Lemaître was interviewed
in Pasadena, shows how little the misunderstandings in the
debate have cleared up.

Given his acceptance of Darwin's theory, it's difficult to
believe that Lemaître would have any interest in the so-called
intelligent design theory of evolution that is currently gaining
support in some political corners today. He flatly disagreed
with any literal interpretation of Genesis. As for the explosion
of New Age thinking, the rise of relativism in humanities
departments, and their concurrent hostility to what is
deemed the negative influence of science in technology, he is
not likely to have been sympathetic to these views either.

Lemaître had a little taste of student unrest himself on the
campus of Louvain in the early 1960s. The Flemish majority
of Belgium's parliament passed a number of language laws,
one of which determined that only Dutch should be used in
schools of the country's northern province, and French in
schools of the southern province. When a movement arose
among Louvain's Flemish faculty to make the university

entirely Dutch-speaking, Lemaître was aroused to make known his opposition. He was elected leader of the French-speaking professors and was rewarded for his efforts when Flemish students smashed the windows of his apartment with rocks. In spite of this turmoil, Lemaître was a popular teacher among the physics students. While his mind could wander in class, sometimes giving his pupils the impression he was as new to the subject he was teaching as they were, he enjoyed meeting outside of class for beer and snacks, where the discussions could be continued in a less formal style.

To the very end, Lemaître was enthusiastic about computers. During the 1950s, when scientific use of computers on campuses was on the rise, Lemaître went so far as to lend Louvain's university rector the funds to purchase the campus's first electronic computer, a Burroughs E101, manufactured in 1957. It looked very like a large, L-shaped office desk, with a typewriter-style keyboard on the desktop. In the next decade, his very last years, Lemaître upgraded to an IBM 1620 and an Elliot 801. On these machines Lemaître taught himself assembler languages and busied himself with numerical calculations of problems in celestial mechanics, which he enjoyed handing on to his students.

Lemaître traveled twice more to the United States, first in 1961, when he was invited to attend the Eleventh General Assembly of the International Astronomical Union at University of California–Berkeley and chaired a seminar on cosmology with several papers presented by other members. In 1962 he visited Berkeley once more, where he wrote his last paper, not on cosmology but on the three-body problem, a perennial subject of celestial mechanics—how to derive an approximate solution to the motion of the moon around the earth considering perturbations in the orbit caused by the sun. Lemaître enjoyed using computers to attack this problem.

Lemaître's involvement in the Church did not diminish in the last years. In 1960 he was made president of the Pontifical Academy of Sciences. Shortly after the death of Pope John XXIII, in the midst of the Second Vatican Council, Lemaître was bemused to find himself nominated to the pontifical commission dealing with birth control.

In late December 1964 Lemaître suffered a mild heart attack while visiting Rome. Upon returning to Louvain, he was hospitalized for a period and put on a strict dietary regimen. While he recovered enough to return to some teaching duties, he soon asked Odon Godart and another colleague to run his daily laborious computer calculations and bring the results to his apartment for him to review. It was clear in the later part of that year that he would not return to his robust self. By the time he was diagnosed with a form of leukemia in the following spring, it was too late to save him.

He was in the Hospital of Saint Peter in Louvain in June of 1966, when on the sixteenth, or perhaps seventeenth, Godart brought him the news that what Penzias and Wilson had discovered the year before was indeed the remnant of the cosmic fireworks Lemaître had always believed would be leftover from his primordial quantum. True, he had expected cosmic rays to be this vestige of the big bang, and true, he had always believed the super-dense state from which the universe had evolved to be a sort of cold, cosmic nucleus, when what Gamow and his team and Peebles and his team had theorized should be a hot, radiation-based big bang. But there was no question now in the minds of most scientists that the universe had indeed evolved from a cosmic origin in time. And Lemaître had been the first one to suggest this, the first to insist that such a state might have left traces of itself in the remoteness of space.

Lemaître died a few days later, on Monday, June 20. George Gamow, often so good-humored about his work, expressed bitterness in the last few years of his life that he and his team did not get any credit for accurately predicting the temperature of the cosmic microwave background radiation. It apparently made no difference to him that Dicke and Peebles revised their paper to give proper credit to him and Alpher and Herman when they realized the $\alpha\beta\gamma$ paper had preceded theirs. Gamow died just two years after Lemaître, in 1968, at only sixty-four years of age. His love for vodka had at last caught up with him. By this time he had long since ceased to contribute any significant papers or research to either cosmology or physics. But he left behind a series of superb popularizations of science that to this day remain classics in print.

The Nobel Prize cannot be awarded posthumously. It was awarded to Penzias and Wilson in 1978, even though they had no idea of what they had discovered at the time. Dicke and Peebles got no recognition. Neither did Alpher and Herman, Gamow's surviving colleagues. On this subject of the credit of scientific discovery, it is perhaps appropriate to close with remarks by Fred Hoyle:

> The microwave background was discovered in 1941. Andrew McKellar, at the Dominion Observatory on Victoria Island, B.C., discovered an excitation temperature of 3K. . . . Factually speaking, this was just as clear cut a discovery of the background as that of 1965, but the world was not in an intellectual position to appreciate it, demonstrating the great importance of sociological factors in assessing the merits of scientific work. Great credit accrues to those who make a scientific discovery when the world is already teetering on the edge of it, whereas the discoverer who is markedly too early scarcely earns a footnote in scientific history.[118]

Just a year after Lemaître died, Irish graduate student Jocelyn Bell at Cambridge University discovered the first neutron star, spinning thirty-six times per second in the heart of the Crab Nebula. The super-dense object that J. Robert Oppenheimer first theoretically predicted in 1939 was now a confirmed reality. Suddenly the collapsed objects that John Archibald Wheeler had christened "black holes" no longer seemed beyond the realm of possibility, and the search for these ultimate collapsed stars, which Oppenheimer and

Snyder modeled using Lemaître's dust solution to the field equations of general relativity, began in earnest. By the late 1990s, the long forgotten cosmological constant, which Lemaître alone considered a vital, intrinsic component of the Einstein equations, was being reconsidered in the light of the new discovery that the universe, as he had long believed, was actually accelerating in its expansion.

It's difficult to imagine any of these current developments displeasing the "little" Belgian priest who before anyone else rigorously sought observational evidence of what the general theory of relativity predicted about the truly dynamic nature of the cosmos. The work that Lemaître began in cosmology back in 1927 is more than a footnote in scientific history, and it continues today.

NOTES

1. Abraham Pais, *Subtle Is the Lord* (Oxford: Oxford University Press, 1982), 240.

2. See, for example, ibid., 444–49; Ronald Clark, *Einstein: The Life and Times* (Avon Discus, 1971), 740–43; Banesh Hoffmann and Helen Dukas, *Albert Einstein: Creator and Rebel* (New American Library, 1972), 186–99.

3. Pais, *Subtle Is the Lord,* 444.

4. See, for example, George Smoot and Keay Davidson, *Wrinkles in Time* (New York: William Morrow, 1993), 54; Odon Godart, "Contributions of Lemâitre to General Relativity," 442, and Jean Eisenstaedt, "Lemâitre and the Schwarzschild Solution," 361, in *Studies in the History of General Relativity,* ed. J. Eisenstaedt, A. J. Kox (Boston: Birkhäuser, 1992).

5. Dominique Lambert, *Un Atome d'Univers* (Brussels: Racine, 2000), 105.

6. See Godart, "Contributions of Lemaître," 437–53.

7. As for Einstein himself, he avoided military service before the out-break of war, though not necessarily because of his age (he was 35 in 1914). As a Swiss citizen he did not have any obligation to the German army, and as an intellectual he found anything to do with the military revolting. Still, relieved though he was at not having to serve, at least one biographer writes that Einstein was humiliated when the Swiss army turned him away because he was flat-footed. But given that he spent the crucial year of 1915 laboring over the general theory of relativity, which inspired the birth of modern cosmology, it's difficult to imagine the course of twentieth-century science had Einstein in any way been prevented from completing his theory.

8. Albrecht Fölsing, *Albert Einstein* (New York: Penguin Books, 1997), 28.

9. Friedrich Gontard, *The Chair of Peter: A History of the Papacy.* Trans. A. J. and E. F. Peeler (New York: Holt, Rinehart and Winston, 1964), 518–19.

10. Helge Kragh, *Quantum Generations* (Princeton, NJ: Princeton University Press, 1999), 3–12.

11. Alfred North Whitehead, *Science and the Modern World* (New York: Macmillan, 1925). As quoted by Kragh, *Quantum Generations.*

12. Lambert, *Un Atome d'Univers,* 25.

13. To this day, ignorance and misunderstanding of Lemaître's background and his specialty continues to color popular accounts of his work in the most slipshod fashion. This is most amusingly on display in Dan Brown's lightweight thriller *Angels and Demons,* where he refers to Lemaître as a "monk" who all along planned to reconcile science and faith specifically by positing the "big bang" theory in 1927. Brown not only errs in assuming the big bang was outlined as such by Lemaître, but he also incorrectly dates Lemaître's version, the primeval atom hypothesis (it was in 1931). He propounds this howler by further stating that Hubble "proved" the theory in 1929 with his famous report on the redshifts of extragalactic nebulae. In fact, Hubble, notoriously cautious to the very end of his life, never claimed any such thing. Not only was there no such theory known as the big bang in 1929, but Hubble suggested in his paper only that the redshifts measured by him and Milton Humason appeared to support nonstatic relativistic models of the cosmos (i.e., expanding universe models), and that there was a direct relation between the distance of nebulae measured and the velocity of their apparent recession.

14. Lambert, *Un Atome d'Univers, 35.*

15. Deprit, *The Big Bang and George Lemaître* (Boston: D. Reidel, 1984), 366.

16. Lambert, 39.

17. Andre Deprit paints a different reaction in his memoir, *Monsignor George Lemaître,* when he writes: "Alas, he challenged an instructor in the classroom on his erroneous solution to a problem in ballistics;

he and his brother were expelled that day from the class, the instructor lodged a complaint, and the commanding officer entered a report stating that Lemaître did not have the attitude expected from a candidate officer. Years after the incident, Lemaître could still not dominate his resentment against the evaluation." Deprit, *The Big Bang and Georges Lemaître,* 363–92.

18. Albert Einstein,"Cosmological Considerations on the General Theory of Relativity" (Sitzungsberichte der Prussichen Akademy de Wissenschaft, 1917), 142–52. As translated in Jeremy Bernstein and Gerald Feinberg, *Cosmological Constants: Papers in Modern Cosmology* (New York: Columbia University Press, 1986), 26.

19. By 1922, when Friedmann wrote his first paper, American astronomer Vesto Slipher had already recorded a handful of redshifts for nebulae he'd photographed at the Lowell Observatory in Flagstaff, Arizona. And Edwin Hubble had also begun to note the redshifts at Mount Wilson Observatory.

20. See, for example, Stanley L. Jaki, *Science and Creation* (Edinburgh: Scottish Academic Press, 1986).

21. Brahmavaivarta Purana. See, for example, in H. Zimmer, *Myths and Symbols in Indian Art and Civilization* (New York: Pantheon Books, 1946), 3–11.

22. Stanley L. Jaki, *Is There a Universe?* The expanded text of three lectures delivered at the University of Liverpool in November 1992. (Liverpool University Press, 1993; New York: Wethersfield Institute).

23. Lambert writes that Eddington was a dull lecturer, which seems surprising given the amount of enthusiasm he worked up for cosmology.

24. In fact, as several biographers of Einstein and historians of science have noted, it wasn't as dramatic as the papers of the time made out. The visibility on Principe was poor during most of the eclipse, and of the three decent plates Eddington was able to take during the period, one of the three disagreed with the predicted figure of star light deflection significantly. For another four decades, the acceptance of the general theory of relativity floated on relatively scant testing, its mathematical elegance, and the simplicity of its principles—until the development of radio and laser tools beginning in the late 1950s and early 1960s introduced new tests that more precisely and decisively confirmed the theory.

25. Lambert, *Un Atome d'Univers,* Brussels: Racine, 2000. See page 69 where he suggests Eddington may have felt an added affinity for Lemaître because he was also unmarried and religious.

26. De Sitter actually gets the credit for first suggesting observed redshifts of nebulae might confirm his model of the cosmos.

27. Lambert, *Un Atome d'Univers,* Brussels: Racine, 2000, 70.

28. J. D. North, *The Measure of the Universe: A History of Modern Cosmology* (New York: Dover 1990), 143.

29. Helge Kragh, *Cosmology and Controversy: The Historical Development of Two Theories of the Universe* (Princeton, NJ: Princeton University Press, 1996), 25–27.

30. Gale E. Christianson, *Edwin Hubble: Mariner of the Nebulae* (New York: Farrar, Straus, and Giroux, 1995). Although Christianson notes that when Hubble was absent from home for any length of time, as he was, for example, during World War II, Grace left her pages blank. Christianson's portrait of the Hubbles is not altogether flattering. Indeed, apart from the the couple's penchant for taking weeks of vacation at the expense of the Mount Wilson Observatory, both Edwin and his wife, apart from his superb astronomical work, could act on occasion like the crassest social climbers, spending a great deal of time with Hollywood celebrities.

31. See, for example, Kragh, *Cosmology and Controversy*, 20–21, and 408, footnote 101.

32. David Bodanis, $E = mc^2$ (New York: Walker & Company, 2000), 215.

33. See, for example, Albrecht Fölsing's biography of Einstein, where Eddington made a joke of pretending not to know who the "third person" in the world could be who understood the equations.

34. Georges Lemaître, "A Homogeneous Universe of Constant Mass and Increasing Radius Accounting for the Radial Velocity of Extra-galactic Nebulae," *Annales de Société Scientifique de Bruxelles* 47 (1927): 49–56.

35. John C. Mather and John Boslough, *The Very First Light* (New York: Basic Books, 1996), 41.

36. Georges Lemaître, "Note on de Sitter's Universe," *Journal of Mathematical Physics* 4 (1925): 188–92.

37. Willem De Sitter, "On Einstein's theory of gravitation and its astronomical consequences," *Monthly Notices of the Royal Astronomical Society* 78 (1917): 3–28.

38. Unaware of Lemaître's work of 1927, Tolman wrote a paper in 1929 considering the connection between the recession of nebulae and the de Sitter effect in much the same way Lemaître had in his 1925 "Note on de Sitter's Universe."

39. From the introduction to Lemaître, "Note on de Sitter's Universe."

40. Edwin Hubble, "A Relation Between Distance and Radial Velocity Among Extra-Galactic Nebulae," *Proceedings of the National Academy of Sciences* 15 (1929): 168–73.

41. George McVittie, Obituary Notice, "Georges Lemaître," *Quarterly Journal of the Royal Astronomical Society* 8 (1967): 294–97.

42. See Georges Lemaître, "Chronique: Recontres avec A. Einstein," *Revue des Questions Scientifique* 129 (1958): 129–32.

43. Albert Einstein, *The Meaning of Relativity* (Princeton, NJ: Princeton University Press, 1945), 126, as quoted in Kragh, *Cosmology and Controversy*, 55.

44. For a popular misreading of this, for example, see Dan Brown's overwrought *Angels and Demons* (New York: Pocket Books, 2000), 69–70. Brown not only gets Lemaître's scientific work wrong, but he also describes Lemaître as a monk and further states that Edwin Hubble proved the big bang was true.

45. P. A. M. Dirac, "The Scientific Work of Georges Lemaître," *Pontificiae Academiae Scientarum Commentarii,* vol. II, no. 11 (1968): 1–20.

46. A. S. Eddington, "The End of the World: From the Standpoint of Mathematical Physics," *Nature* 127, (1931): 447–53.

47. Georges Lemaître, "The beginning of the world from the point of view of quantum theory," *Nature* 127 (1931): 706.

48. Robert H. Dicke, *Gravitation and the Universe* (Philadelphia: American Philosophical Society, 1970).

49. Lemaître's derivation of Hubble's law two years before Hubble published it himself and the fact that he derived an expansion rate very close to Hubble's does support the contention.

50. "Salvation Without Belief in Jonah's Whale," *Literary Digest* 115 (March 11, 1933): 23.

51. Kragh, *Cosmology and Controversy,* 55, and corresponding note on 408.

52. There is a great deal of discussion going on at present over the "dark energy" and whether it in fact can be represented by the cosmological constant or should, as Kirshner argues, be represented by some other force. One problem with the constant is that it must be set to a far greater number to match the current expansion rates than seems credible. What remains significant, though, is that Lemaître believed from the outset that some other force existed to propel the expansion.

53. Gödel looked at the possibility of a rotating spherical cosmos—one that, among other things, would allow time travel into the past. Lemaître narrowly missed meeting Gödel when the latter was too ill to come to Notre Dame during the 1938 conference Lemaître spoke at during his own semester there as a visiting lecturer on cosmology.

54. George Gamow, *My World Line* (New York: Viking Press, 1970), 45.

55. Fred Hoyle, "Final Remarks," in Guido Chincarini, et al., eds., *Observational Cosmology* (San Francisco: Astronomical Society of the Pacific, 1993), 695.

56. Ralph Alpher and Robert Herman, *Genesis of the Big Bang* (Oxford: Oxford University Press, 2001), 70.

57. It's become part of scientific legend to point out that Gamow added Hans Bethe's name to this historic paper as a sort of pun, spinning their names, Alpher, Bethe, and Gamow, off the first letters of the Greek alphabet—alpha, beta and gamma. This sort of playfulness was typical of Gamow, and Bethe, who was actually one of the reviewers for the article when it was submitted, enjoyed the joke, all the more so as he was enthusiastic about its thesis.

58. See Kragh, p. 134 on the work of Australian astrophysicist Andrew McKellar in 1940.

59. Ralph A. Alpher and Robert Herman, "Early Work on 'Big-Bang' Cosmology and the Cosmic Blackbody Radiation," in B. Bertotti, et al., eds. *Modern Cosmology in Retrospect* (Cambridge: Cambridge University Press, 1990), 129–58.

60. Odon Godart and Michael Heller, *The Cosmology of Lemaître* (Tucson: Pachart, 1985), 133.

61. See Kragh, *Cosmology and Controversy,* 57–58.

62. Astrophysicist John Gribbin argues that had both men lived into the 1970s, they would have shared the Nobel Prize that went to Penzias and Wilson. See John Gribbin, *In Search of the Big Bang* (New York: Bantam, 1986), 193.

63. Although it is far from accurate to say that the rest of the physics establishment was convinced. Eddington's results from the 1919 expedition, for example, were second-guessed and criticized for years. See Jean Eisenstaedt, *Einstein and the History of General Relativity,* ed. Don Howard and John Stachel, Vol. 1, Boston: Birkhäuser, 1989, 277–92.

64. Peter G. Bergmann, *Introduction to the Theory of Relativity* (New York: Prentice Hall, 1942), 211.

65. J. Robert Oppenheimer, "Einstein," *Review of Modern Physics* 28 (1956): 1–2. As quoted in Jean Eisenstaedt, "The Low Water Mark of General Relativity," in *Einstein and the History of General Relativity,* Vol. 1 (1989), Don Howard, John Stachel, eds. 283.

66. In fact, Sir Arthur Eddington also applied the term "Bang" to the primeval explosion when Lemaître first proposed it (see Kragh, *Cosmology and Controversy*). It's interesting that in this context, while no enthusiast for the steady state theory, Eddington also felt compelled to deride the idea of a temporal beginning to the evolution of the cosmos.

67. The curator of Hoyle's papers told this author his widow does recall their trip—and that they got on well. No more details were provided, other than that they "agreed to disagree" about the big bang and the steady state. Hoyle himself left a more personal recollection of their relationship, which is described in chapter 11.

68. Kragh, *Cosmology and Controversy*, 239.

69. For example, John North's otherwise excellent *Norton History of Astronomy and Cosmology* (New York: W. W. Norton, 1995), 526.

70. See Kragh, *Cosmology and Controversy*, 255.

71. Peter Michelmore, *Einstein: Profile of the Man* (New York: Dodd, Mead, 1962), 253.

72. Fred Hoyle, "Final Remarks," in Chincarini, et al., eds., *Observational Cosmology*, 694–95.

73. Robert Kirshner, *The Extravagant Universe* (Princeton, NJ: Princeton University Press, 2002), 75. Kirshner writes that he's not sure Einstein ever said this, as the quote is attributed to the ever-flamboyant Gamow from his autobiography *My World Line,* and Gamow was known to embellish a story for dramatic effect. Gamow claimed Einstein told him this during one of their conversations together. On page 44 of his "informal" biography, he writes of his former teacher Friedmann's exploring expanding universe models after setting $\Lambda = 0.$ "Thus Einstein's original gravity equation was correct, and changing it was a mistake. Much later, when I was discussing cosmological problems with Einstein, he remarked that the introduction of the

cosmological term was the biggest blunder he ever made in his life. But this 'blunder,' rejected by Einstein, is still sometimes used by cosmologists even today and the cosmological constant denoted by the Greek letter Λ still rears its ugly head again and again and again."

74. Pierre Speziali, ed. *Albert Einstein—Michele Besso: Correspondence, 1903–1955* (Paris: Hermann, 1979), 68.

75. Hugo von Seeliger and Carl von Neumann come to mind. In the mid 1890s they submitted a slightly altered gravitational force in order to get around the problem. Seeliger had also suggested this as a way to deal with the problem of the advance of the perihelion of Mercury. See Kragh, *Cosmology and Controversy,* 6.

76. Jeremy Bernstein and Gerald Feinberg, *Cosmological Constants* (New York: Columbia University Press), 9.

77. Had Einstein accepted the indication that a cosmic solution of his equations entailed an expanding universe in 1917, Gribbin argues, he could have made the single greatest prediction in the history of science before Hubble's 1929 discovery.

78. Kragh, *Cosmology and Controversy,* 53.

79. Hermann Weyl, *Space, Time, Matter* (London: Methuen, 1922), 297.

80. Arthur S. Eddington, *The Expanding Universe* (Cambridge: Cambridge University Press, 1933), 24.

81. See A. Vibert Douglas, *The Life of Arthur Stanley Eddington* (London: Thomas Nelson and Sons, 1956), 163.

82. Lemaître's correspondence with Einstein and the paper were obtained from the Albert Einstein Archives, the Jewish National and University Library of the Hebrew University of Jerusalem. His paper was published in volume 2 of *Albert Einstein: Philosopher-Scientist,* edited by Paul Arthur Schilpp (London: Cambridge University Press, 1949).

83. Allan Sandage, "Current Problems in the Extra Galactic Distance Scale," *Astrophysical Journal* 127 (1958): 513–26. From 525.

84. Alan Guth, *The Inflationary Universe* (Reading, MA: Addison Wesley, 1997), 175.

85. It should be noted that in 1923 Hermann Weyl also suggested the same thing before Lemaître, but he did so indirectly, and without tying the strange recession effect of matter in the de Sitter universe to any actual observations.

86. Richard C. Tolman, "Effect of Inhomogeneity in Cosmological Models," *Proceedings of the National Academy of Sciences* 20 (1934): 169–76.

87. See, for example, a recent paper by Adam G. Riess, et al., from the *Astrophysical Journal,* 2004, referred to the author in an e-mail sent by Professor Steve Carlip at UC Davis, which puts the cosmic relation of pressure and density fairly close to Λ.

88. Karl Schwarzschild, "Über das Gravitationsfeld einer kugel aus inkompressibler Flüssigkeit nach der Einsteinschen Theorie," in *Sitzungsberichte der Königlich Preussischen Akademie der Wissenschaften* (Berlin, 1916), 424–34.

89. Jean Eisenstaedt, "Lemaître and the Schwarzschild Solution," in *The Attraction of Gravitation*, Vol. 5, J. Earman, M. Janssen, J. D. Norton, eds. (Boston: Birkhäuser, 1993), 362.

90. Ibid., 366.

91. Georges Lemaître, "L'Univers en Expansion," *Publication du Laboratoire d'Astronomie et de Géodesie de l'Universite de Louvain* 9 (1932): 171–205. Later published in *Annales de Société Scientifique de Bruxelles* 53 (1933): 51–85.

92. Tolman, "Effect of Inhomogeneity," 169–76.

93. Hermann Bondi, "Spherically Symmetrical Models in General Relativity," Royal Astronomical Society. *Monthly Notices* 107 (1947): 410–25.

94. J. L. Synge, "On the Expansion or Contraction of a Symmetrical Cloud under the Influence of Gravity," National Academy of Sciences. *Proceedings* 20 (1934): 635–40.

95. J. Robert Oppenheimer and H. Snyder, "On Continued Gravitational Contraction," *Physical Review* 56 (1939): 455–59.

96. J. Robert Oppenheimer and G. M. Volkoff, "On Massive Neutron Cores," *Physical Review* 55 (1939): 374–81.

97. Eisenstaedt, "Lemaître and the Schwarzschild Solution," 370.

98. Charles W. Misner, Kip S. Thorne, and John Archibald Wheeler, *Gravitation* (New York: Freeman, 1973), 620.

99. Eisenstaedt, "Lemaître and the Schwarzchild Solution," 372.

100. Ibid.

101. Ibid.

102. Ibid., 373.

103. Ernan McMullin, "How should cosmology relate to theology?" in A. R. Peacocke, ed., *The Sciences and Theology in the Twentieth Century* (Notre Dame, IN: University of Notre Dame Press, 1981), 53–54.

104. John Cornwell's *Hitler's Pope* and David Kertzer's *Pope Against the Jews* are two examples arousing recent controversy.

105. Paul Johnson, *A History of Christianity* (New York: Atheneum, 1976), 503.

106. "Un Ora," *Acta Apostolicae Sedes—Commentarium Officiale,* 44 (1952): 31–43.

107. *Time* 58 (Dec. 3, 1951): 75–77.

108. George Gamow, "The role of turbulence in the evolution of the universe," *Physical Review* 86 (1952): 251.

109. Hoyle, "Final Remarks," in Chincarini, et al., eds. *Observational Cosmology,* 694–95.

110. Lambert, *Un Atom d'Univers,* 280.

111. George Sylvester Viereck, "What Life Means to Einstein," *Saturday Evening Post,* October 26, 1929.

112. "Salvation Without Belief in Jonah's Tale," *Literary Digest* 115 (March 11, 1933): 23.

113. Fred Hoyle, *Man and Materialism* (New York: Harper and Brothers, 1956), 218.

114. Kragh, *Cosmology and Controversy,* 253.

115. Fred Hoyle, "Frontiers in Cosmology," in S. K. Biswas, D. C. V. Mallik, and C. V. Vishveshwara, eds., *Cosmic Perspectives: Essays Dedicated to the Memory of M. K. V. Bappu* (Cambridge: Cambridge University Press, 1989), 101.

116. William B. Bonnor, *The Mystery of the Expanding Universe* (New York: Macmillan, 1964), 117.

117. Lemaître's Solvay talk, "The primeval atom hypothesis and the problem of the clusters of galaxies," was reprinted in R. Stoops, ed., *La Structure et l'Evolution de l'Univers* (Brussels: Coudenberg, 1958), 1–32.

118. Hoyle, "Final Remarks," in Chincarini, et al., eds., *Observational Cosmology,* 695.

GLOSSARY

Big bang: term first coined by Fred Hoyle during his 1950 radio addresses on the nature of the universe. Hoyle used it to refer to relativistic models of the universe, such as Lemaître's and Gamow's, that evolved from a super-dense state or singularity.

Cepheid variable: a type of variable star named after Delta Cepheus, whose brightness was seen to change over time. Henrietta Swann Leavitt discovered in 1912 that an approximately linear relationship existed between the period of a Cepheid's pulsation and its luminosity. The longer the period of its pulsation, the greater was the star's luminosity. Once the distance to nearby Cepheids was determined, they became useful measuring sticks for establishing the distances to other galaxies where they were discovered. In 1925 Edwin Hubble discovered a Cepheid in M31, the Andromeda Nebula, which enabled him to estimate that it was eight hundred thousand light-years from earth, and convinced most astronomers that Andromeda was not part of the Milky Way, but in fact a galaxy in its own right.

Cosmic microwave background radiation: the low-level hum of radio waves detectable from every direction in space, measuring about three degrees Kelvin, above absolute zero. It was predicted by Gamow, Alpher, and Herman in 1948 as a consequence of the initial super-hot stage of the universe after the big bang. As the universe expanded, the radiation has cooled over the billions of years to the attenuated microwaves now detectable. Initially discovered accidentally by Penzias and Wilson in 1965, it constituted the most solid evidence in favor of the big bang theory.

Cosmological constant: also known as the lambda term Λ, which Einstein used to maintain a static equilibrium in his initial cosmological solution to the general relativity field equations in 1917. Without the term, he realized his spherical model of the universe would collapse in on itself, like a balloon losing all of its air. Although he introduced the term on the left side of his field equations, essentially to keep the balloon inflated, or "prop up" the geometry of space-time, Lemaître later reinterpreted Λ, as a measure of negative pressure, which he incorporated on the right side of the equations dealing with stress-energy and its effect on the curvature of space-time. The constant has been understood in these terms ever since.

De Sitter effect: described by Sir Arthur Stanley Eddington, the redshifting and apparent recession of particles in the de Sitter model of the universe.

De Sitter model: solution to Einstein's field equations by Dutch astronomer Willem de Sitter, proposed in 1917, which showed that (1) Einstein's cosmological solution was not the only one, and (2) one could posit an infinite, spatially flat model of space-time essentially devoid of matter. A curious effect of this model was the apparent recession of particles of matter introduced into space-time. Lemaître later reinterpreted de Sitter's solution to show that it was, in fact, the first limited model of expanding space. Fred Hoyle would base his steady state model on de Sitter's, as would Alan Guth in 1979 when he developed his theory of inflation.

Doppler effect: light from a source moving relative to an observer will change in wavelength. Christian Doppler first proposed this effect in terms of pitch for sound waves in 1841, but modern astronomers apply it to electromagnetic waves, corresponding to an observed change in color. Stars or galaxies with spectra shifted to the blue, or shorter, wavelength end of the spectrum are considered approaching the point of view of the earth. Stars and nebulae with spectra shifted to the red, or longer, wavelengths are considered moving away.

Dust solution: a solution to Einstein's field equations that Lemaître developed in 1932 in order to revisit Schwarzschild's solution. Instead of a uniform sphere of fluid, whose pressure would go to infinity and thus establish a singularity at the Schwarzschild limit, Lemaître suggested using a sphere of dust, of zero pressure, and was able to show that in fact the singularity at Schwarzschild's limit was only apparent; one could theoretically model a sphere whose radius collapsed to a singularity at radius $r = 0$. J. Robert Oppenheimer used Lemaître's solution in 1939 to posit the first model of what became known as a black hole.

Eddington model: also known as Lemaître-Eddington model, the first model proposed by Lemaître in which the universe expands from an initial, static Einstein state and accelerates into de Sitter's empty model. Even after Lemaître posited his primeval atom theory of cosmic origin, many astronomers and cosmologists, including Eddington, preferred this model because it did not posit a temporal origin.

Einstein static state: The universe model proposed by Einstein in his 1917 paper, which exists as a static, four-dimensional sphere, held in equilibrium against collapse by the cosmological constant, finite in terms of matter, but unbounded in terms of space-time.

Expansion of the universe: the expansion of cosmic space suggested by Einstein's field equations of general relativity. Alexander Friedmann showed in 1922 that such an expansion was a natural consequence of the field equations, but Einstein disagreed and essentially ignored Friedmann (who died in 1925) until Lemaître independently redeveloped his model along more rigorous lines.

General theory of relativity: Einstein's theory of universal gravitation, which describes gravity in geometric terms, as a property of space-time rather than as a force. The geometry of space, or the curvature of space-time, is determined by the presence of mass and energy. Among its first known consequences, the theory predicted that light would bend in the presence of massive objects and that gravity would also slow clocks in the field of massive objects. Einstein early on realized the theory's importance for cosmology; as a geometric theory, general relativity allowed Einstein in theory to model the entire universe as the totality of all existing matter. This opened up an entire new field of cosmology as an exact science in the twentieth century.

Homogeneity: for simplicity, an attribute assumed for early cosmological models, such as those of Lemaître and Einstein—the assumption that matter is distributed evenly throughout the volume of the universe.

Hubble constant: denoted by the symbol H_0, the Hubble constant is the rate at which the universe is estimated to be expanding. Astronomers measure it in terms of kilometers per second, per megaparsec. It is currently estimated to be roughly 70 kilometers per second per megaparsec of space—or 20 kilometers per second per million light-years of space.

Hubble's law: the velocity of a galaxy in recession, as determined by its redshift, is proportional to its distance. The farther away the galaxy is estimated to be, the faster its velocity of recession. It can be written as $v = H_0 \times r$, where v = velocity, H_0 is the Hubble constant, and r is the radius or distance to a galaxy. Lemaître derived Hubble's law from his existing data in his 1927 paper, two years before Hubble proposed it in his paper on the recession of the galaxies.

Hubble time: estimate for the age of the universe based on the reciprocal of Hubble's constant. At the time Lemaître proposed his primeval atom theory, the Hubble time was only estimated to be about two billion years. Later revisions of Hubble's distance estimates increased the value upward to four billion years in 1948 and upward to ten billion by the time Allan Sandage took over Hubble's position at Mount Palomar. Current Hubble time is still believed to be between ten and twenty billion years.

Isotropy: for simplicity, this attribute was also assumed by Einstein, Lemaître, and other cosmologists—that the universe appears the same in all directions. To a very high degree, measurements of the cosmic microwave background radiation bear this out.

Lambda: see cosmological constant.

Lemaître's model: a relativistic model of the universe that begins from a cold dense kernel or "primeval atom," expands rapidly with positive cosmological constant, into a stagnating or "hesitating" state for eons during which the stars and galaxies evolve, before expanding again in an accelerating rate.

Primeval atom (*l'hypothese de l'atom primitif):* the name Lemaître gave to his theory of the universe initiating from a hyper-dense sphere of cold matter, disintegrating via radioactivity into an expanding universe. His theory was also more accurately described as "the cosmic egg."

Redshift: the displacement of spectral lines of a measured astronomical object (e.g., star, galaxy) toward the red end of the spectrum, indicating distance and velocity of recession. Astronomers interpreted the displacement as due to the Doppler effect, the change in wavelength characteristic of motion on the part of a source relative to an observer.

Schwarzschild solution: the first complete solution to Einstein's field equations of general relativity, proposed by German astronomer Karl Schwarzschild shortly before he died in 1916. Schwarzschild assumed that for a uniform, fluid sphere, a singularity would occur as the sphere's radius shrank to $r = 2GM/c^2$, where G is the gravitational constant, c is the speed of light, and M is the mass of the sphere. This radius is now considered the event horizon of a black hole, the radius at which light cannot escape from a star. Lemaître later showed that any singularity at this radius was only apparent, and that under different coordinates and assumptions, the singularity would occur at $r = 0$.

Singularity: a mathematical point in the equations of general relativity where one or more elements run to an infinite value, in effect making the equations no longer viable. A black hole is an example where the curvature of space-time itself rises to infinity. From an aesthetic point of view, Einstein considered the singularity a drawback of his theory. He urged Lemaître to explore solutions to the field equations that might avoid it.

Space-time: the four-dimensional coordinate system, or domain, of general relativity. Events in space-time are characterized by the three spatial dimensions—x, y, and z—as well as time t, the fourth dimension.

Special theory of relativity: Einstein's theory combines two principles— that the speed of light is constant, irrespective of the velocity of its source, and that the laws of physics hold universally for all inertial frames of reference, meaning coordinate systems at rest or moving with a constant velocity. Two consequences of the theory are (1) that observers in different reference frames moving relative to one another will record different results for the time of events in a third frame; and (2) that energy is equal to mass multiplied by the speed of light squared.

Spectroscopy: The study of spectra. In the nineteenth century it was discovered that spectra taken of the sun and the stars exhibited dark absorption lines corresponding to the chemical elements they were composed of. These lines corresponded to the characteristic absorption lines of chemicals in the laboratory, giving scientists a new tool to determine the chemical composition of stars and gaseous nebulae.

Spectrum (spectra): a recording, usually on a plate, of the distribution of electromagnetic waves (light) according to wavelength. By taking spectra of stars and galaxies, astronomers can determine chemical composition, redshift, distance, and apparent velocity.

Steady state theory: a cosmological theory postulated by Fred Hoyle, Thomas Gold, and Herman Bondi in Britain in 1948, reasserting the Aristotelian premise that the universe is in essence without change. Hoyle, Bondi, and Gold argued against a temporal, big bang origin, suggesting that while the universe was expanding, this occurred in a natural state similar to that suggested by de Sitter's universe, but with matter being constantly created in the form of hydrogen atoms in empty space to sustain the existence of the cosmos in perpetua. The theory enjoyed some support in Britain through the 1950s, but in the early 1960s the discovery of quasars and other developments in astronomy strengthened the contention that the universe in the distant past was far different than at present. The discovery of the cosmic microwave background radiation all but discredited the theory, although Hoyle periodically tried to modify it in the ensuing decades.

Type Ia supernovae: violent stellar explosions, signaling the utter demise of a star, which take place when a white dwarf bleeds enough mass from a companion star to pass a certain mass limit, approximately 1.4 times the mass of the sun (also called Chandrasekhar's Limit after the Indian-American astrophysicist Subrahmanyan Chandrasekhar), at which point it explodes. The luminance of such supernovae is so great that they can be brighter than the entire galaxies in which they are photographed. The consistent level of luminosity of Type Ia supernovae have made them an even more reliable distance indicator than Cepheids for determining extra-galactic distances.

Type II supernovae: explosions of stars of masses at least eight times that of the sun. Having reached a state where all nuclear fusion has reduced the star's core to iron and heavier elements, the star can no longer sustain fusion to provide the energy necessary to counteract the force of gravitational collapse. The core contracts into a neutron core or black hole, while the outer layers of lighter matter are blown off into space at thousands of kilometers per second.

BIBLIOGRAPHY

For further reading about the life and work of Georges Lemaître, a few books are not in print, but most are still in circulation at your local library.

Helge Kragh's *Cosmology and Controversy* (1996) is a superb and somewhat technical overview of twentieth-century cosmology, with a chapter on Lemaître and his influence.

Dominique Lambert's recent biography of Lemaître, *Un Atome d'Univers* (2000), as of this writing available only in French, is an exhaustive account of Lemaître's family as well as his life and work. This includes a great many of Lemaître's noncosmological interests.

Odon Godart's book, *Cosmology of Lemaître* (1985), written with Michael Heller, is also an excellent overview, with Godart's own personal recollections of Lemaître. It is no longer in print, but it should be available at most university libraries.

Finally Lemaître's own book, *The Primeval Atom* (1950), long out of print, can also be found at most university libraries. Lemaître was an evocative writer, and even the technical sections of this book are recounted with panache.

Alpher, Ralph A., and Robert Herman. *Genesis of the Big Bang.* Oxford: Oxford University Press, 2001.

———. "Early work on 'big-bang' cosmology and the cosmic blackbody radiation." In B. Bertotti, et al., eds. *Modern Cosmology in Retrospect.* Cambridge: Cambridge University Press, 1990.

Barrow, John D. *The Origin of the Universe.* New York: Basic Books, 1994.

Berger, André, ed. *The Big Bang and Georges Lemaître: Proceedings of a Symposium in Honour of G. Lemaître Fifty Years after His Initiation of Big-Bang Cosmology.* Dordrecht: D. Reidel, 1984.

Bergmann, Peter G. *Introduction to the Theory of Relativity.* New York: Prentice Hall, 1942.

Bernstein, Jeremy, and Gerald Feinberg. *Cosmological Constants: Papers in Modern Cosmology.* New York: Columbia University Press, 1986.

Bodanis, David. $E = mc^2$. New York: Walker, 2000.

Bondi, Hermann. "Spherically Symmetrical Models in General Relativity." Royal Astronomical Society. *Monthly Notices* 107 (1947).

Bonnor, William B. *The Mystery of the Expanding Universe.* New York: Macmillan, 1964.

Bibliography

Brian, Denis. *Einstein: A Life.* New York: John Wiley & Sons, 1996.

Brown, Dan. *Angels and Demons.* New York: Pocket Books, 2000.

Chincarini, Guido, et al., eds. *Observational Cosmology: International Symposium, Held in Milano, Italy, 21–25 September 1992.* San Francisco: Astronomical Society of the Pacific, 1993.

Clark, Ronald. *Einstein: The Life and Times.* New York: Avon Discus, 1971.

Christianson, Gale E. *Edwin Hubble: Mariner of the Nebulae.* New York: Farrar, Straus, Giroux, 1995.

Danielson, Dennis Richard. *Book of the Cosmos: Imagining the Universe from Heraclitus to Hawking.* New York: Helix Books, 2000.

Deprit, Andre. "Monsignor Georges Lemaître" in *The Big Bang and George Lemaître.* Boston: D. Reidel, 1984.

De Sitter, Willem. "On Einstein's Theory of Gravitation and Its Astronomical Consequences." *Monthly Notices of the Royal Astronomical Society* 78 (1917).

Dicke, Robert H. *Gravitation and the Universe.* Philadelphia: American Philosophical Society, 1970.

Dirac, P. A. M. "The Scientific Work of Georges Lemaître." *Pontificiae Academiae Scientarum Commentarii.* Vol. II, no. 11 (1968).

Douglas, A. Vibert. *The Life of Arthur Stanley Eddington*. London: Thomas Nelson and Sons, 1956.

Dukas, Helen and Banesh Hoffman. *Albert Einstein: Creator and Rebel*. New York: New American Library, 1972.

Earman, John, Michel Janssen, and John D. Norton, eds. *The Attraction of Gravitation: New Studies in the History of General Relativity*. Einstein Studies, Vol. 5. Boston: Birkhäuser, 1993.

Eddington, Arthur Stanley. *The Expanding Universe*. Cambridge: Cambridge University Press, 1933.

———. Eddington, Arthur Stanley. "The End of the World: From the Standpoint of Mathematical Physics." *Nature* 127 (1931).

———. *Mathematical Theory of Relativity*. Cambridge: Cambridge University Press, 1924.

———. *Space, Time, and Gravitation: An Outline of the General Relativity Theory*. Cambridge: Cambridge University Press, 1920.

Eisenstaedt, Jean. "Lemaître and the Schwarzschild Solution," in *Studies in the History of General Relativity*, ed. J. Eisenstaedt, A. J. Kox. Boston: Birkhäuser, 1992.

Ferguson, Kitty. *Measuring the Universe: Our Historic Quest to Chart the Horizons of Space and Time*. New York: Walker, 1999.

Ferris, Timothy. *The Whole Shebang: A State-of-the-Universe(s) Report*. New York: Touchstone, 1997.

Fölsing, Albrecht. *Albert Einstein: A Biography.* New York: Penguin Books, 1997.

Gamow, George. *My World Line: An Informal Autobiography.* New York: Viking, 1970.

———. *Creation of the Universe.* New York: Viking, 1961.

———. "The Role of Turbulence in the Evolution of the Universe." *Physical Review* 86 (1952).

Godart, Odon. "Contributions of Lemaître to General Relativity." In *Studies in the History of General Relativity,* ed. J. Eisenstaedt, A. J. Kox. Boston: Birkhäuser, 1992.

Godart, Odon, and Michael Heller. *The Cosmology of Lemaître.* Tucson: Pachart, 1985.

Gontard, Friedrich. *The Chair of Peter: A History of the Papacy.* Trans. A. J. and E. F. Peeler. New York: Holt, Rinehart & Winston, 1964.

Gribbin, John. *In Search of the Big Bang: Quantum Physics and Cosmology.* New York: Bantam Books, 1986.

Guth, Alan H. *The Inflationary Universe: The Quest for a New Theory of Cosmic Origins.* Reading, MA: Addison-Wesley, 1997.

Harrison, Edward Robert. *Cosmology: The Science of the Universe.* 2nd ed. Cambridge: Cambridge University Press, 2000.

Hawking, Stephen W., Kip S. Thorne, et al. *The Future of Spacetime.* Intro. Richard Price. New York: W. W. Norton, 2002.

Heller, Michael. *Lemaître: Big Bang and the Quantum Universe.* Tucson: Pachart, 1996.

Hoyle, Fred. *Man and Materialism.* New York: Harper and Brothers, 1956.

———. *The Nature of the Universe.* New York: Harper & Row, 1950.

———. "Frontiers in Cosmology," in *Cosmic Perspectives: Essays Dedicated to the Memory of M. K. V. Bappu,* ed. S.K. Biswas, D.C. V. Mallik, C.V. Vishveshwara. Cambridge: Cambridge University Press, 1989.

Howard, Don and John Stachel, eds. *Einstein and the History of General Relativity.* Boston: Birkhäuser, 1989.

Hubble, Edwin. "A Relation Between Distance and Radial Velocity Among Extra-Galactic Nebulae." *Proceedings of the National Academy of Sciences* 15 (1929).

Jaki, Stanley L. *Science and Creation: From Eternal Cycles to an Oscillating Universe.* Edinburgh: Scottish Academic Press, 1986.

———. *Is There a Universe?* New York: Wethersfield Institute, 1993.

Johnson, Paul. *A History of Christianity.* New York: Atheneum, 1976.

Kirshner, Robert P. *The Extravagant Universe: Exploding Stars, Dark Energy, and the Accelerating Cosmos.* Princeton, NJ: Princeton University Press, 2002.

Kragh, Helge. *Quantum Generations: A History of Physics in the Twentieth Century.* Princeton, NJ: Princeton University Press, 1999.

———. *Cosmology and Controversy: The Historical Development of Two Theories of the Universe.* Princeton, NJ: Princeton University Press, 1996.

Lambert, Dominique. *Un Atome D'univers: La Vie et L'Oeuvre de Georges Lemaître.* Brussels: Racine, 2000.

Lemaître, Georges. "Chronique: Recontres avec A. Einstein." *Revue des Questions Scientifique* 129 (1958).

———. *The Primeval Atom: An Essay on Cosmogony.* Trans. Betty H. and Serge A. Korff. Toronto: D. Van Nostrand, 1950.

———. "The beginning of the world from the point of view of quantum theory." *Nature* 127 (1931).

———. "A Homogeneous Universe of Constant Mass and Increasing Radius Accounting for the Radial Velocity of Extragalactic Nebulae." *Annales de Société Scientifique de Bruxelles* 47 (1927).

———. "L'Univers en Expansion," in *Publication du Laboratoire d'Astronomie et de Géodesie de l'Universite de Louvain* 9 (1932).

———. "Note on de Sitter's Universe." *Journal of Mathematical Physics* 4 (1925).

[Lemaître, Georges] Pietro Salviucci. *L'Academie Pontificale des Sciences en Memoire de Son Second President Georges Lemaître a L'Occasion du*

Cinquieme Anniversaire de Sa Mort. Rome: Pontificia Academia Scientiarum, 1972.

Levy, David H., ed. *Scientific American Book of the Cosmos.* New York: St. Martin's Press, 2000.

Lightman, Alan P. *Ancient Light: Our Changing View of the Universe.* Cambridge: Harvard University Press, 1991.

Lightman, Alan, and Roberta Brawer. *Origins: The Lives and Worlds of Modern Cosmologists.* Cambridge: Harvard University Press, 1990.

Literary Digest (no author listed). "Salvation without Belief in Jonah's Tale." *Literary Digest* 115 Number 10 (1933).

Livio, Mario. *Accelerating Universe: Infinite Expansion, the Cosmological Constant, and the Beauty of the Cosmos.* New York: Wiley, 2000.

Mather, John C., and John Boslough. *The Very First Light: The True Inside Story of the Scientific Journey Back to the Dawn of the Universe.* New York: Basic Books, 1996.

Michelmore, Peter. *Einstein: Profile of the Man.* New York: Dodd, Mead, 1962.

Misner, Charles W., Kip S. Thorne, and John Archibald Wheeler. *Gravitation.* New York: Freeman, 1973.

Munitz, Milton K. *Theories of the Universe:* From Babylonian Myth to Modern Science. New York: Free Press, 1957.

North, J. D. [John David]. *The Measure of the Universe: A History of Modern Cosmology.* New York: Dover, 1990 [1965].

——, ed. *Norton History of Astronomy and Cosmology.* New York: W. W. Norton, 1995.

Oppenheimer, J. Robert, and H. Snyder. "On Continued Gravitational Contraction." *Physical Review* 56 (1939).

Oppenheimer, J. Robert, and G. M. Volkoff. "On Massive Neutron Cores." *Physical Review* 55 (1939).

Pais, Abraham. *"Subtle Is the Lord—": The Science and the Life of Albert Einstein.* Oxford: Oxford University Press, 1982.

Peacocke, A. R. [Arthur Robert], ed. *The Sciences and Theology in the Twentieth Century.* Notre Dame, IN: University of Notre Dame Press, 1981.

Rees, Martin. *Just Six Numbers: The Deep Forces that Shape the Universe.* New York: Basic Books, 2000.

Sandage, Allan. "Current Problems in the Extra Galactic Distance Scale." *Astrophysical Journal* 127 (1958).

Schutz, Bernard F. *A First Course in General Relativity.* Cambridge: Cambridge University Press, 1990.

Schilpp, Paul Arthur, ed. *Albert Einstein: Philosopher-Scientist.* London: Cambridge University Press, 1949.

Seife, Charles. *Alpha and Omega: The Search for the Beginning and End of the Universe.* New York: Viking, 2003.

Shapley, Harlow. *A Source Book in Astronomy, 1900–1950.* Cambridge: Harvard University Press, 1979 [1960].

Silk, Joseph. *The Big Bang.* 3rd ed. New York: W. H. Freeman, 2001.

Smoot, George, and Keay Davidson. *Wrinkles in Time.* New York: William Morrow, 1993.

Speziali, Pierre, ed. *Albert Einstein—Michele Besso: Correspondence, 1903–1955.* Paris: Hermann, 1979.

Synge, J. L. "On the Expansion or Contraction of a Symmetrical Cloud under the Influence of Gravity." National Academy of Sciences. *Proceedings* 20 (1934).

Tauber, Gerald. *Albert Einstein's Theory of General Relativity: Sixty Years of Its influence on Man and the Universe.* New York: Crown, 1979.

Tolman, Richard C. *Relativity, Thermodynamics, and Cosmology.* Oxford: Clarendon Press, 1934.

———. "Effect of Inhomogeneity in Cosmological Models." *Proceedings of the National Academy of Sciences* 20 (1934): 169–76.

Viereck, George Sylvester. "What Life Means to Einstein." *Saturday Evening Post,* October 26, 1929.

Bibliography

Weinberg, Steven. *The First Three Minutes: A Modern View of the Origin of the Universe.* New York: Bantam Books, 1977.

Weyl, Hermann. *Space, Time, Matter.* Trans. Henry L. Bros. London: Methuen, 1922.

Zimmer, H. Myths and Symbols in Indian Art and Civilization. New York: Pantheon Books, 1946.

INDEX

Index